Simone Mascotto

Preparation and Novel Characterization Ways of Mesoporous Metal Oxides

Simone Mascotto

Preparation and Novel Characterization Ways of Mesoporous Metal Oxides

sol-gel synthesis of mesoporous metal oxides and their characterization combining small-angle scattering and gas physisorption methods

Südwestdeutscher Verlag für Hochschulschriften

Imprint
Any brand names and product names mentioned in this book are subject to trademark, brand or patent protection and are trademarks or registered trademarks of their respective holders. The use of brand names, product names, common names, trade names, product descriptions etc. even without a particular marking in this work is in no way to be construed to mean that such names may be regarded as unrestricted in respect of trademark and brand protection legislation and could thus be used by anyone.

Publisher:
Südwestdeutscher Verlag für Hochschulschriften
is a trademark of
Dodo Books Indian Ocean Ltd., member of the OmniScriptum S.R.L Publishing group
str. A.Russo 15, of. 61, Chisinau-2068, Republic of Moldova Europe
Printed at: see last page
ISBN: 978-3-8381-1990-8

Zugl. / Approved by: Giessen, Justus Liebig Universitaet Giessen, Diss., 2009

Copyright © Simone Mascotto
Copyright © 2010 Dodo Books Indian Ocean Ltd., member of the OmniScriptum S.R.L Publishing group

When I was a small boy, I used to walk that section of London around the British Museum; and one day, I came across a shop, which had a notice over the window, which said "Philosophical Instruments". Even as a boy I knew something about Philosophy, but I couldn´t imagine what "Philosophical Instruments" could be. So I went up to the window, and there displayed, were chronometers, slide rules, scales and all kind of what we now call "Scientific Instruments". Because Science used to be called Natural Philosophy. Because as Aristotle says: "The beginning of Philosophy is wonder. Philosophy is man´s expression of curiosity about everything, his attempt to make sense of world primarily through his intellect", that he says is faculty for thinking. And thinking, of course, is a word used in many ways and is a very weighed word for most people, but I use the word thinking in a very precise way. By thinking, as distinct from feeling or emoting or sensing, I mean the manipulation of symbols whether they´d be words, whether they´d be numbers or whether they´d be other such signs let´s say triangles, squares, circles, astrological signs or whatever.

Alan W. Watts - Speech about Thinking

Contents

Introduction		**1**
1	**Physical methods**	**5**
1.1	Gas physisorption	5
	1.1.1 General aspects	5
	1.1.2 Determination of the porous features	7
1.2	Small-angle x-ray scattering (SAXS)	11
	1.2.1 Principles of scattering theory	11
	1.2.2 Ordered systems	16
	1.2.3 Disordered systems	18
	1.2.4 2D low-angle transmission SAXS	21
1.3	In-situ SAXS/SANS-physisorption	23
1.4	Wide-angle x-ray scattering (WAXS)	25
1.5	Electron microscopy	26
1.6	Atomic force microscopy (AFM)	27
2	**Synthesis of mesoporous metal oxides using new amphiphilic block copolymers**	**29**
2.1	Introduction	29
2.2	Nanocasting by liquid crystal templating of block copolymers	32
2.3	Synthesis and characterization	34
	2.3.1 Nanocasted SiO_2 powders	35
	2.3.2 Hierarchical SiO_2 systems	41
	2.3.3 SiO_2 Films	45
	2.3.4 TiO_2 Films	54
2.4	Mesoporous TiO_2 thin films as photovoltaic devices	62
	2.4.1 Current state of research	62

	2.4.2	Characteristics of the photovoltaic cell	65
	2.4.3	Testing of dye sensitized solar cells	65
2.5	Summary	. . . .	68

3 Microporosity determination of hierarchical mesoporous SiO$_2$ by in-situ SANS   69

3.1	Introduction	69
3.2	Physisorption analysis	72
3.3	In-situ SANS data	74
3.4	Analysis of the in-situ SANS data by the CLD concept	76
3.5	Analysis of the Bragg peak intensity	81
3.6	General aspects of pore filling behavior	84
3.7	Self-aggregation study of block copolymers by micropore analysis	85
3.8	Summary	87

4 Vapor physisorption on hierarchical SiO$_2$ by in-situ SAXS/SANS   89

4.1	Introduction	89
4.2	Vapor physisorption analyses	90
4.3	Adsorption in-situ SAXS/SANS data	92
4.4	Desorption in-situ SANS data	96
4.5	Summary	100

5 Vapor physisorption on PMO by in-situ SAXS/SANS   103

5.1	Introduction	103
5.2	Materials investigated	106
5.3	Vapor physisorption analyses	107
5.4	In-situ SAXS/SANS data	110
5.5	Analysis of the in-situ SAXS curves	112
5.6	Summary	117

Conclusion   119

A Appendix   123

A.1	Synthesis strategies	123
	A.1.1 Nanocasted silica powders	123
	A.1.2 Nanocasted hierarchical silica powders	123

	A.1.3 Mesoporous silica films	123
	A.1.4 Mesoporous titania films	124
	A.1.5 Preparation of the DSSC devices	125
A.2	Analytical methods	127
A.3	In-situ SAXS/SANS-physisorption	128
	A.3.1 In-situ SAXS-physisorption	128
	A.3.2 In-situ SANS-physisorption	129
A.4	Mathematical appendix	131
	A.4.1 Expressions of $F(s)$ and $S(s)$ for hard-spheres system in the PY model	131
	A.4.2 Expressions of $F(s)$ and $S(s)$ for hard-discs system in the PY model	132
	A.4.3 Calculation of the $\widetilde{\phi}_{micro}$ from the Bragg peak intensity	133
A.5	Discussions on micropore analysis	134
	A.5.1 Monolayer formation of adsorbate in hierarchical SiO_2	134
	A.5.2 Volume fraction calculation by means of the Bragg peak analysis	135
A.6	Additional physisorption analyses	137
	A.6.1 N_2 physisorption on different batch of PIB-IL	137
	A.6.2 Langmuir plot of CH_2Br_2 physisorption on PMOs	138
A.7	List of Chemicals	138

Acronyms and Abbreviations   139

Symbols   141

Bibliography   143

List of Figures   158

List of Tables   159

VIII

Introduction

Scientists are all along fascinated by the high degree of complexity and miniaturization found in natural materials. Nature is indeed the greatest school for materials science providing a multiplicity of composites, architectures, systems and functions [1–4]. Highly elaborated natural materials assemblies are for example crustacean carapaces, mollusc shell, bone or teeth tissue. New materials and systems provided by the man must in the future aim at higher level of sophistication, be environmental friendly and save energy. Thus, to improve and reach higher performances, they have to be prepared following a biomimetic approach, which requires both an understanding of the basic building principles of living organisms and a study of the physico-chemical properties of the synthesis processes.

Even though high complex and aesthetic natural structures pass well beyond current accomplishments realized in materials science, advances in the field called "organized matter chemistry" [5] show promising man-made materials, as illustrated in numerous publications of the last decade [6–9]. Such systems are generated for instance by emulating self-assembly principles occurring in biological systems like formation of molecular crystals, liposomes, micelles and bilayered membranes [10–14]. Thus, the study of complex molecules assemblies resulting from the fine tuning of intramolecular interactions, also known as supramolecular chemistry, allows the controlled design of new materials with complex and original structures. In fact, liquid crystal phases formed from surfactants or amphiphilic block copolymers can serve as moulds to obtain mesoporous metal oxide materials with a variety of different morphologies, controlled porosity and textures [15–17]. The intrinsic physico-chemical properties of the amorphous/crystalline materials, together with the high internal surface area and large pore volume, make these systems extremely attractive for technical applications. Such advanced materials are expected to be employed in fields as optics, catalysis, sensors and separa-

Introduction

tion [18]. According to the classification made by IUPAC [19], porous materials can be arranged in three main categories depending on the pore size (diameter, d), in micro- ($d < 2$ nm), meso- (2 nm $< d < 50$ nm) and macroporous solids ($d > 50$ nm).

Although supramolecular chemistry provides attractive and fascinating opportunities for the establishment of smart nanoporous materials, the conscious exploitation of its potential could be comprehended, till some years ago, only to a certain extent, due to the lack of appropriate characterization methods. Thus, the need to create improved and higher-performing materials for the advancement of science and technology, has to indispensably pass through the development of quantitative physical methods for the understanding of the materials' chemical and physical properties. With particular regard to porous materials, small-angle scattering and gas physisorption represent, since many years, the most reliable methods for the structural investigation of these composites. Nevertheless, these techniques alone can not completely explain complex phenomena like capillary condensation and desorption mechanisms of gases in mesopores, or finely elucidate the structure of sophisticated porous textures. In the light of this, a novel analytical procedure consisting in the combination of gas physisorption with small-angle scattering methods (SAXS/SANS), i.e. in-situ SAXS/SANS-physisorption, is found to be an elegant solution for the porosity and structure determination of these systems. This approach consists in the acquisition of scattering curves of the material at different adsorption/desorption steps, being able, by tuning the experimental conditions, e.g. adsorbable gases and temperature, to probe the spatial distribution of the pores, their connectivity and accessibility and to directly elucidate gas condensation and desorption mechanisms.

In the present work the preparation of novel nanoporous textures and their characterization by means of in-situ SAXS/SANS-physisorption methodologies are faced.

In CHAPTER 2 the self-assembly, i.e. micellization, of a new class of amphiphilic block copolymers with high hydrophobic contrast is deeply investigated through the generation of porous metal oxides, e.g. silica and titania, in form of powders, films and hierarchical systems. Besides the bare templating behavior studies, more interesting systems like mesoporous TiO_2 thin films are also tested as dye sensitized solar cell devices. Moreover, the preparation of complex hierarchical architectures covers a remarkable part in this chapter, since it represents one of the future trends for the development of high performance materials. Inspired by similar biological structures, these hierarchical textures are characterized by an elevated degree of miniaturization and sophistication. Owning to this, being standard analytical methods often inadequate for their structural investigation, a significant part of this work is devoted to the characterization of these hierarchical mesostructures by in-situ SAXS/SANS-physisorption.

In CHAPTER 3 in-situ SANS-nitrogen physisorption studies at 77 K will be presented in order to characterize the microporosity of hierarchical silicas. The analysis of the scattering curves will present alternative solutions with respect to the bare nitrogen physisorption to quantify the micropores, providing also direct information on their spatial distributions.

In CHAPTER 4 in-situ SAXS/SANS-physisorption experiments on hierarchical silicas will be performed also employing organic vapors at room temperature conditions. The chance to study the physisorption behavior using organic molecules of different sizes will give important insights into the understanding of the pore accessibility and connectivity and provide direct evidences of adsorption/desorption mechanisms in such mesoporous materials.

Further investigations by means of in-situ SAXS/SANS-physisorption of organic vapors will be addressed in CHAPTER 5 to periodica mesoporuos organosilica systems. In this case, besides the porosity studies particular attention is given to the elucidation of the spatial distribution of the organic moieties in these hybrid organic-inorganic materials.

Chapter 1

Physical methods

This chapter gives an introduction on the main analytical methods and data evaluation approaches used in this work.

1.1 Gas physisorption

1.1.1 General aspects

Gas physisorption represents one of the most popular methods for the characterization of porous materials, since it allows to assess a wide range of pore sizes (from 0.35 to 100 nm) including the complete range of micropores and mesopores. The phenomenon of physisorption, or more precisely of the gas *adsorption*, occurs when an adsorbable gas (the *adsorptive*) is brought on the surface of a solid (the *adsorbent*). The gas in the adsorbed state is known also as *adsorbate*. The intramolecular interactions involved during this process are weak dipole-dipole and Van der Waals forces. The counterpart of the adsorption, namely the *desorption*, denotes the decrease in the amount of the adsorbed substance. The relation, at constant temperature, between the quantity adsorbed and the equilibrium pressure of the gas is known as *physisorption isotherm* (Fig. 1.1). The physisorption process in porous materials can be devided into several levels. The *micropore filling* regards the primary filling of pore space, since the micropores are the smallest pores (< 2 nm). Subsequently, in the *monolayer adsorption* all the adsorbed molecules are in contact with the surface layer of the adsorbent. The *multilayer adsorption* process is referred to the adsorption region which contains more than one layer of molecules, and not all the adsorbed molcules are in contact

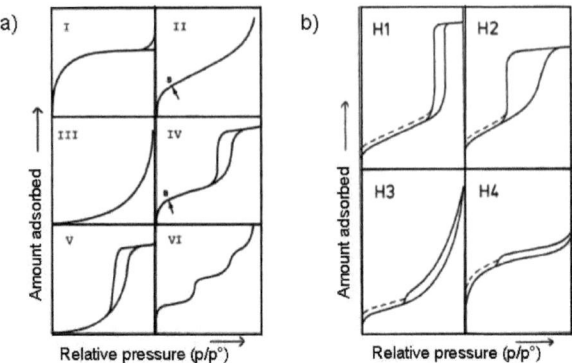

Figure 1.1: (a) Classification of physisorption isotherms for different kinds of adsorbent. (b) Different types of hysteresis in mesoporous materials [20].

with the surface of the adsorbent. The residual pore space left after multilayer adsorption, is filled with fluid at the liquid state. This phenomenon is denoted as *capillary condensation* and is often accompanied by hysteresis.

The majority of physisorption isotherms can be classified into six types showed in Fig. 1.1a depending on the porosity of the adsorbent and on the adsorbent-adsorbate interactions. The most common classes of isotherms are the *Type I* and *Type IV* isotherms. The former are given by microporous solids. The filling of the micropores takes place at small partial pressures and, since no more pores are present, a plateau is formed. Within this work particular relevance is given to the analysis of *Type IV* isotherms. These isotherms are typical of a mesoporous system and are characterized by several features. First of all the hysteresis loop, which is associated with the capillary condensation in mesopores. The arrow in Fig. 1.1a indicates the point corresponding to the stage where the monolayer formation and the micropore filling is accomplished and the multilayer formation is about to begin. At large $p/p°$ values, all the pores being filled, the formation of a plateau is expected. The hysteresis loop is surely the most characterizing feature of type IV isotherms, even though it is not clearly understood. The most realistic explanation to this effect takes into account the thermodynamics of the fluid condensed in the pores. The hysteresis is then associated with the thermodynamic metastability of low and high density phases of the adsorbate [21]. Depending on the mesostructure of the materials, different kinds of hysteresis can be formed (Fig. 1.1b). The most important ones in the context of this work are type H1 and H2. The first one, possessing parallel adsorption and desorption branches, is referred to systems with well defined porous structure and pore geometry, while the second one is characteristic for disordered materials with broad distribution of pore size and not well defined pore shape.

1.1.2 Determination of the porous features

The big benefit of the gas physisorption method is that it allows a profound characterization of the porous materials allowing the determination of surface area, pore volume, pore size and pore size distribution just in one measurement. The surface area of the adsorbent may be calculated by the amount of adsorbate needed to cover the surface with a complete monolayer of molecules (*monolayer capacity*). The application of the Brunauer-Emmet-Teller (BET) method [22] has become the most widely used standard procedure for the determination of the surface area of porous materials. In this context, the following relation can be obtained

$$\frac{p}{n_a(p^\circ - p)} = \frac{1}{n_m C} + \frac{C-1}{n_m C}\frac{p}{p^\circ} \qquad (1.1)$$

where n_a is the amount of adsorbate, expressed in moles, adsorbed at the relative pressure p/p° and n_m is the monolayer capacity. The term C is related exponentially to the enthalpy of adsorption in the monolayer and gives an indication of the order of magnitude of adsorbate-adsorbent interaction energy. In order to apply the BET equation a linear relation between $p/n_a(p^\circ - p)$ and p/p°, i.e. the BET plot, is needed. This range is restricted to the first part of the isotherm, namely in the range $0.05 \leq p/p^\circ \leq 0.3$. For the calculation of the BET surface area (S$_{BET}$), the knowledge of the molecular cross sectional area (a_m) of the adsorbate is necessary, which in the case of N$_2$ at 77 K is 16.2 Å^2. Then

$$S_{BET} = n_m a_m N_A$$

where the N_A is the Avogadro constant. It is noteworthy to point out that the BET equation is strictly not applicable in the case of a microporous adsorbent and in the case of mesoporous materials having pore widths < 4 nm [23, 24]. This latter finding is due to probable pore filling at pressures very close to the pressure range of monolayer-multilayer formation which leads to a significant overestimation of the monolayer capacity in the case of the BET analysis.

The presence of a distinct plateau in the sorption isotherms allows the calculation of the total specific pore volume (V_p). This value can be obtained by converting the amount of adsorbed gas at the saturation pressure $p°$ (m_{ads}) into liquid volume, assuming that the density of the adsorbate (ρ_{ads}) is equal to the bulk liquid density (Gurvich law) [21]

$$V_p = \frac{m_{ads}}{\rho_{ads}}. \tag{1.2}$$

The amount of adsorbed gas can be obtained through the relationship

$$m_{ads} = MW_{ads}\, n_{ads} = MW_{ads}\frac{V_{ads}}{22400[cm^3 mol^{-1}]}$$

where MW_{ads} and n_{ads} are the molecular weight and the moles of adsorbate at the saturation pressure, respectively. Since a mole of gas occupies a volume of 22400 cm^3, in ideal condition, the moles of adsorbate can be calculated. Finally, V_{ads} is the volume of adsorbate at the saturation pressure, which is obtained directly by the sorption isotherm.

The capillary condensation step in isotherms of mesoporous materials is associated with a shift of the vapor-liquid coexistence in pores, compared to the bulk fluid. A fluid confined in a pore condenses at a pressure lower than the saturation pressure at a given temperature. The condensation pressure depends mainly on pore size and shape. Thus, the adsorption isotherm contains direct information about the pore size distribution in the sample. The capillary condensation can be described on the basis of the Kelvin equation in terms of macroscopic magnitudes like the surface tension of the bulk fluid γ, the densities of the coexistent liquid ρ_l and gas ρ_g, $\Delta\rho = (\rho_l - \rho_g)$, the contact angle θ of the liquid meniscus against the pore wall. Assuming a cylindrical shape of the pores, the modified version of the Kelvin equation [23–25] is

$$\ln\frac{p}{p°} = -\frac{2\gamma\cos\theta}{RT\Delta\rho(r_p - t_c)} \tag{1.3}$$

where R is the universal gas constant, r_p the pore radius and t_c the thickness of an adsorbed multilayer film, which is formed prior to pore condensation. The most used approach which adopts the modified Kelvin equation for the pore size distribution (PSD) analysis is the Barret-Joyner-Halenda (BJH) method [26].

Even though it is widely applied, this model can only be employed for cylindrical pore geometries, becoming highly imprecise in the prediction of the PSD for pores narrower than 4-5 nm [27,28]. In many cases the BJH approach results inaccurate because macroscopic models (Kelvin equation) assume that the pore condensation consists in a gas-liquid phase transition in the core of the pore only between two *homogeneous*, bulk-like gas and liquid phases [21]. In contrast, microscopic approaches like the non-local density functional theory (NLDFT) [29,30] suggest that a fluid confined to a single pore can exist with two possible density profiles corresponding to *inhomogeneous* gas and liquid configurations in the pore. The NLDFT model correctly describes the local fluid structure near the pore surface, and quantitatively agrees with molecular simulations [30]. To calculate the pore size distribution of spherical cavities, which are the most investigated within this work, the experimental isotherm is represented as a combination of theoretical isotherms in individual pores

$$N_{exp}(p/p^\circ) = \int_{D_{min}}^{D_{max}} N_V^{ex}(D_{in}, p/p^\circ) \varphi_V(D_{in})\, dD_{in}. \quad (1.4)$$

Here, $N_V^{ex}(D_{in}, p/p^\circ)$ is a kernel of the theoretical isotherms in pores of different diameter and $\varphi_V(D_{in})$ is the PSD function. Both members are function of D_{in}, which represents the "internal" pore diameter, i.e. the free pore diameter immediately before the capillary condensation step [31].

The NLDFT model predicts that in the region of capillary condensation hysteresis there are two configurations of density distribution: a low density *vapor-like* state and a high density *liquid-like* state. An *equilibrium transition*, (p_e in Fig. 1.2a) occurs when the thermodynamical conditions of the vapor-like and liquid-like states are equal. Above the equilibrium transition pressure the vapor-like states on the adsorption branch become *metastable*. When the limit of stability of these metastable states is reached (p_{sv}), the fluid condenses spontaneously and the adsorption branch terminates (*spinodal condensation*) [32].

On the contrary, the desorption process of a mesoporous material can also be used for the pore size determination, but its interpretation is more involved, since the draining behavior depends not only on the type of porous texture geometry but also on the experimental conditions, i.e. temperature and adsorptive. For materials which present a highly ordered structure (H1 type isotherms) like the cylindrical MCM-41 or the gyroidal KIT-6 silicas, the desorption pressure falls near the equilibrium transition region ($p_e < p_d < p_c$ in Fig. 1.2a). This

1. Physical methods

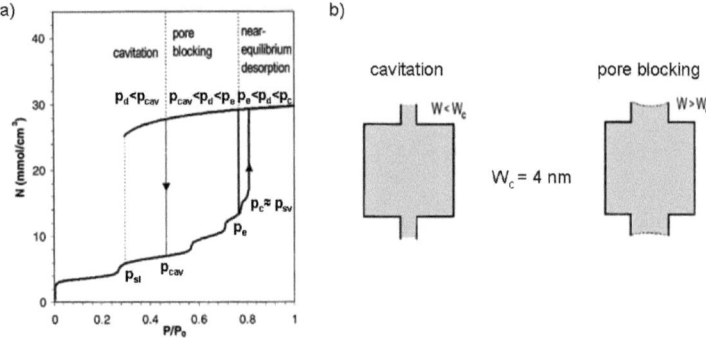

Figure 1.2: (a) Isotherm simulation by the NLDFT approach showing the theoretical pressures of liquid-like spinodal, equilibrium and vapor-like spinodal, denoted as p_{sl}, p_e and p_{sv}. Furthermore the three regimes of evaporation are classified with respect to the pressure of desorption p_d. The pressure of spontaneous condensation and cavitation is denoted as p_c and p_{cav} respectively [33]. (b) Representation of ink-bottle pore for cavitation and pore-blocking emptying mechanism; with W the neck diameter and W_c the critical neck diameter [34].

phenomenon occurs, because in these materials the cavities have direct access to the vapor phase, establishing a liquid-to-vapor equilibrium transition [33]. In this case, since a stable transition occurs, the desorption behavior can be described through macroscopic models and the BJH can be used for the PSD determination. In the case of less ordered materials (H2 type isotherms) or cage-like structures, which are objects of study in this work, the cavities are blocked and have direct access to the vapor phase only through a narrower neck (ink-bottle pore geometry, Fig. 1.2b). In this scenario the desorption is delayed and passes into the plateau region, corresponding to the reduction of density of condensed liquid in the pores. Here the desorption process is controlled by the pore geometry and precisely by the size of the pore neck. If the pore neck is bigger than 4 nm (Fig. 1.2b), condition valid only in the case of nitrogen at 77 K, the desorption is driven by the so-called *pore-blocking* effect ($p_{cav} < p_d < p_e$). In this model the pore core remains filled until the narrower neck empties first at lower vapor pressure. Hence, in a network of ink-bottle pores, evaporation of the capillary condensate is obstructed by the pore neck [34,35]. If the neck diameter is smaller than 4 nm the desorption occurs via *cavitation*. In cavitation regime the desorption pressure is independent by the pore neck size and is determined by the spontaneous evaporation of the capillary condensate (p_{cav}) in the core of the pore. The spontaneous evaporation occurs in proximity of the liquid-like spinodal, p_{sl} (*spinodal evaporation*), where the metastable state of condensed liquid reaches the limit of stability. Thus, the pore body can empty by diffusion while the pore neck remains filled [31,35]. In this latter case, it is well documented in literature [36–38] that the desorption pressure is independent by the material pore structure and determined primarily by the kind of adsorptive and its physisorption temperature. Several works [33,34] also demonstrated that the change of the experimental conditions leads to a transition between cavitation and pore blocking effect.

In the light of these findings two important aspects can be remarked. The first is that for ink-bottle pore systems the BJH method is inappropriate for the PSD determination, since the desorption behavior can be explained only through the formation of metastable liquid states. Secondly, since the draining of the pores can be strongly affected by the experimental conditions, a reliable PSD is obtained only by the use of the NLDFT model for the adsorption branch.

1.2 Small-angle x-ray scattering (SAXS)

1.2.1 Principles of scattering theory

The structural and morphological features of a porous material can be exhaustively studied also through small-angle scattering experiments. This technique allows to investigate structures from 1 to 100 nm and thus is particularly adequate for the analysis of mesoporous materials. Although within this work both photons and neutrons were used as scattering source, the following treatment will be dealt with x-rays only, since the theory holds true for neutron scattering as well, without significant modifications.

Diffraction is generally produced by the interference of waves scattered by a body. In the case of x-rays striking the body, every electron becomes source of scattered waves. In small-angle range ($0 < \theta < 5°$) it has been assumed that all the scattering waves are coherent, neglecting incoherent scattering, thus implying that the respective amplitudes from all the scattering centers are added. Summation

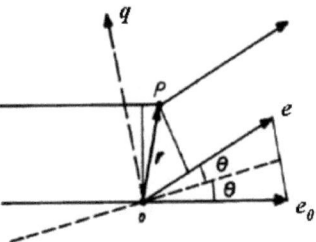

Figure 1.3: Scattering representation by two point centers [39].

can be replaced by integration over the whole volume V irradiated by the incident beam

$$A(\mathbf{q}) = \iiint dV \rho_e(\mathbf{r}) \exp(-i\mathbf{rq}) \qquad (1.5)$$
$$A(\mathbf{q}) = \mathscr{F}(\rho(\mathbf{r})).$$

In this sense the amplitude A is the Fourier transform $\mathscr{F}$ of the electron density distribution within the object, where $\rho_e(\mathbf{r})$ is the electron density, defined as the number of electron per unit volume (cm^3), and $\exp(-i\mathbf{rq})$ represents a single scattered wave in the complex form. The path difference of a point P (Fig. 1.3), specified by the vector $\mathbf{r}$, is seen as $-\mathbf{r}(\mathbf{e}-\mathbf{e}_o) = -2\mathbf{r}\sin\theta$. The modulus of the scattering vector $\mathbf{q}$ can be expressed thus in the form $q = (4\pi/\lambda)\sin\theta$, where λ is the wavelength of the incident beam.

The scattering intensity $I(\mathbf{q})$ can be obtained by Eqn. 1.5 as the absolute square of the amplitude by using the complex conjugate A^*

$$\begin{aligned}
I(\mathbf{q}) &= |A(\mathbf{q})|^2 = A(\mathbf{q})\,A^*(\mathbf{q}) \\
&= \iiint dV_1 \rho_e(\mathbf{r}_1)\exp(-i\mathbf{r}_1\mathbf{q}) \iiint dV_2 \rho_e(\mathbf{r}_2)\exp(i\mathbf{r}_2\mathbf{q}) \\
&= \iiint dV \iiint dV_1 \rho_e(\mathbf{r}_1)\rho_e(\mathbf{r}_2)\exp(-i\mathbf{rq}) \qquad (1.6) \\
&= \iiint dV\, P(\mathbf{r})\exp(-i\mathbf{rq}). \qquad (1.7)
\end{aligned}$$

The autocorrelation function $P(\mathbf{r})$, also known as Patterson function, expresses the density of two points with distances $\mathbf{r}_1$ and $\mathbf{r}_2$ from the origin. From Eqn. 1.7 the measurable scattering intensity is then the Fourier transform of the autocorrelation function $I(\mathbf{q}) = \mathscr{F}(P(\mathbf{r}))$. The reciprocity between real and reciprocal space dependence, namely between an object owning a certain $\rho_e(\mathbf{r})$ and its scattering intensity $I(\mathbf{q})$, is exemplified in the following scheme:

$$\begin{array}{ccc} \rho_e(\mathbf{r}) & \xleftrightarrow{\mathscr{F}} & A(\mathbf{q}) \\ *^2 \downarrow & & \downarrow ||^2 \\ P(\mathbf{r}) & \xleftrightarrow{\mathscr{F}} & I(\mathbf{q}) \end{array} \qquad (1.8)$$

In most cases, the investigated systems are assumed to be statistically isotropic and without long range order. The first restriction allows to consider the distribution $P(r)$ depending only on the values of the distance r. The second instead, allows to replace, at high r values, the electron density (ρ_e) with the mean value $\bar{\rho}_e$. Thus, throughout the total volume is convenient to use the density fluctuation $\eta = \rho_e - \bar{\rho}_e$ instead of the density itself

$$\bar{\eta}^2 = \widetilde{(\rho_e - \bar{\rho}_e)}^2 = P(r) - VP(r) = V\gamma(r) \qquad (1.9)$$

where $\gamma(r)$ is the so-called correlation function, for which the properties $\gamma(0) = \langle \eta^2 \rangle$ and $\gamma \to 0$ are valid for larger r. Moreover, the correlation function for an isotropic system is found to be the inverse of the Fourier transform of the intensity

$$V\gamma(\mathbf{r}) = \frac{1}{2\pi} \int_0^\infty \mathbf{q}^2 d\mathbf{q}\, I(\mathbf{q}) \frac{\sin \mathbf{qr}}{\mathbf{qr}} \qquad (1.10)$$

where the term $\frac{\sin qr}{qr}$ represents the averaged phase factor $\langle \exp(-irq) \rangle$.
The quantity

$$Q = \int_0^\infty \mathbf{q}^2 d\mathbf{q}\, I(\mathbf{q}) \qquad (1.11)$$

is also called "invariant" of the system.

The scattering of a *single particle system* can be discussed by the use of $\gamma(r)$, assuming a constant difference of electron density $(\Delta \rho_e)^2$

$$\gamma(r) = (\Delta \rho_e)^2 \gamma_0(r)\,;\quad \gamma_0(0) = 1\,;\quad \gamma_0(r \geq D) = 0 \qquad (1.12)$$

where $\gamma_0(r)$ gives information about the the geometry of the particle. Imaging the particle shifted of a vector $\mathbf{r}$ (Fig. 1.4a), the volume $\hat{V}$ in common with the particle and its "ghost", contains all the points which give a contribution to $\gamma_0(r)$. The particle looks cut into rods of different length l, which are termed "chords". The group of chords for all directions may be considered using a distribution function $g(l)$, i.e. chord length distribution function, [40, 41] such that $g(l)dl$ is

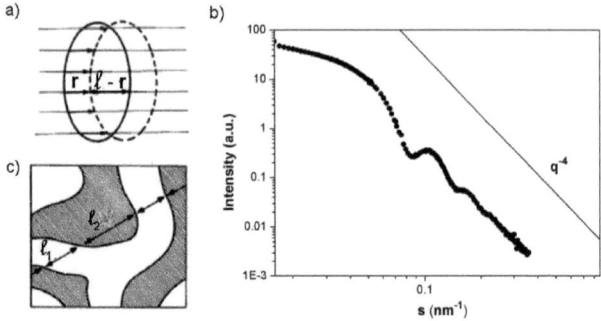

Figure 1.4: Chord length for a single particle (a) and a two-phase system (c) [39]. Typical scattering pattern for a polydispersed spherical single particle system (b).

the probability that a chord randomly chosen is of length between l and $(l+dl)$. It can be seen in Fig. 1.4a that for each chord with $l > r$ a piece $(l-r)$ is contained in $\hat{V}(r)$, thus

$$\gamma_0(r) = \frac{1}{l_p}\int_r^D (l-r)g(l)dr\,; \quad \text{with} \quad l_p = \int_0^D lg(l)dl. \tag{1.13}$$

By differentiation it can be found

$$\frac{d\gamma_0(r)}{dr} = -\frac{1}{l_p}\int_r^D g(l)\,dl\,; \quad \frac{d^2\gamma_0(r)}{dr^2} = \frac{1}{l_p}g(r). \tag{1.14}$$

In this sense the chord length distribution function $g(l)$ is directly referred to the correlation function $\gamma_0(r)$ and can be proper used for the representation of a particle with respect to its scattering pattern.

The parameters characteristic of the dimensions of the particle, like volume and particle radius can be determined by the central part of the scattering pattern (Fig. 1.4b). The final slope of the curve reveals instead the fine structure of the particle, which can be expressed by the behavior of $\gamma_0(r)$ at small r (real space). Expanding the correlation function into a power series

$$\gamma_0(r) = 1 - ar + br^2 + \ldots\,; \quad \text{with} \quad a = \frac{1}{l_p} = \frac{S}{4V} \tag{1.15}$$

the first parameter a is obtained by Eqn. 1.14 and it depends on the surface of the particle. The parameter b is closely related to the initial values of the chord length distribution, since Eqn. 1.14 implies $b = g(0)/2l_p$. In general this factor is equal to zero, except in the case of particles bearing edges or corners. Hence, it might be said that the parameter b, and thus $g(0)$, measures the "angularity" of the system [41]. Substituting then the $\gamma_0(r)$ in Eqn. 1.12, the scattering intensity can be obtained by Eqn. 1.10 opportunely converted. Thus

$$I(q) \rightarrow (\Delta\rho)^2 V \frac{8\pi}{l_p} \frac{1}{q^4} = (\Delta\rho)^2 \frac{2\pi}{q^4} S. \qquad (1.16)$$

This fourth power law, also called *Porod law* [42], holding true at large scattering vector values, is a general statement being also valid for densely packed systems and non-particulate structures. The asymptotic value of $I(q)$ means that a well defined internal surface it exists

$$\lim_{q \to +\infty} I(q)q^4 = \frac{Q\,S}{\pi\,V}. \qquad (1.17)$$

Besides the characterization of diluted particles systems, the small-angle scattering is particularly attractive for the analysis of *two-phase systems* like colloidal dispersions or porous materials. An ideal two-phase system consists of two different substances of constant electron density each and different volume fraction ϕ_1 and ϕ_2. The sample is characterized then by an averaged electron density $\bar{\rho}_e$ all over the volume and respective mean square density fluctuations $\langle \eta^2 \rangle$

$$\bar{\rho}_e = \phi_1 \rho_{e_1} + \phi_2 \rho_{e_2}\,; \qquad \langle \eta^2 \rangle \equiv \langle (\rho_e - \bar{\rho}_e)^2 \rangle = (\rho_{e_1} - \rho_{e_2})^2 \phi_1 \phi_2. \qquad (1.18)$$

The latter magnitude, $\langle \eta^2 \rangle$, is directly dependent from the scattering intensity through the "invariant" Q of the system

$$Q \equiv \int_0^\infty q^2 I(q) dq = V \langle \eta^2 \rangle \, 2\pi^2 = V(\rho_{e_1} - \rho_{e_2})^2 \phi_1 \phi_2 \, 2\pi^2. \qquad (1.19)$$

This dependence is also known as the Babinet principle of reciprocity. The intensity of the whole system is now expressed by

$$I(q) = V\phi_1\phi_2(\Delta\rho)^2 \int_0^\infty 4\pi r^2 \gamma_0(r) \frac{\sin qr}{qr} \qquad (1.20)$$

and, mathematically speaking, differs from the single particle case only by the volume fraction factors (ϕ_1, ϕ_2). The resulting scattering plot must then show the

1. Physical methods

same general features: bell shaped central part and final slope (Fig. 1.5a). In this case, however, the meaning of $\gamma_0(r)$ is less direct than for a single particle, describing more the averaged surrounding of the phase border than the geometrical properties. Nevertheless, the concepts of size and shape can still be applied, but in a different way than for particles.

Regarding the relation between the final slope and the intensity function, the single particle treatment is still valid but a modification is needed, since the surface in this case belongs both to phase "one" and to phase "two". Considering the chord distribution (Fig. 1.4c), the system will cut out alternating chords l_1 and l_2 from the two regions. The respective Porod lengths have to be proportional then with the volume fractions: $l_{p_1} : l_{p_2} = \phi_1 : \phi_2$. Thus from Eqn. 1.15

$$l_p = l_{p_1}\phi_2 = l_{p_2}\phi_1 = 4\phi_1(1-\phi_1)\frac{V}{S}; \quad \text{or:} \quad \frac{1}{l_p} = \frac{1}{l_{p_1}} + \frac{1}{l_{p_2}}. \qquad (1.21)$$

In the limiting case of a diluted system l_p becomes equal to l_{p_1}, without any contradiction to the Babinet principle, if the particles alone are taken into account. The final slope expression is essentially the same as the one in Eqn. 1.17

$$\lim_{q\to+\infty} I(q)q^4 = \frac{Q\,S}{\phi_1\phi_2\pi\,V}. \qquad (1.22)$$

It is noteworthy to mention that in the following data treatment sections, as well as in the discussion of the results, the scattering vector **s** will be adopted. This is related to **q** by the relation: $q = 2\pi\,s$.

1.2.2 Ordered systems

In the case of ordered two-phase systems, objects possessing an electron density ρ_{e_1} are displaced in a matrix (ρ_{e_2}) with a defined structure similar to those employed in the classical crystallography. The scattering intensity can be represented in the form

$$I(\mathbf{s}) = |F(\mathbf{s})|^2 |S(\mathbf{s})|^2 \qquad (1.23)$$

where $F(s)$ is the form factor and $S(s)$ the lattice factor (structure factor) of the objects. Within this work the synthesis of mesoporous materials possessing well defined spherical mesopores allowed the use of specific structural models for the correct prediction of the spatial distribution and morphology of the pores. Analytical calculations of the scattering intensity on similar systems [43, 44] have

1.2.2. Ordered systems

been performed assuming a polydispersed hard-sphere model in the Percus-Yevick (PY) approximation [45,46]. Moreover, in our case, since the equation expressing the scattering intensity is quite complicated, it has been assumed that the mesopores' position is independent of their size [47–49]. Thus the scattering intensity is given by

$$I(s) = k[\langle |F(s)|^2 \rangle + |\langle F(s) \rangle|^2 (\langle |S(s)|^2 \rangle - 1)] \quad (1.24)$$

where k is a scaling factor and the $\langle \rangle$ brackets represent the number-average of the spheres. The polydispersity of the mesopores can be described using a Gaussian distribution for the number distribution of radii

$$h(r) = \frac{1}{\sigma_R \sqrt{2\pi}} \exp\left[-\frac{(R - \langle R \rangle)^2}{2\sigma_R^2}\right] \quad (1.25)$$

with the average radius $\langle R \rangle$ and the variance σ_R as parameters. The form factor of a sphere of radius R is given by

$$F(R, s) = \left(\frac{R}{s}\right)^{\frac{3}{2}} J_{\frac{3}{2}}(R, s) \quad (1.26)$$

where $J_{\frac{3}{2}}(R, s)$ is a Bessel function of the first kind of the order $\frac{3}{2}$. The structure factor $S(s)$ is obtained through the Percus-Yevick approach with the parameters R_{PY} and η_{PY} [49,50]

$$S(s) = \left[\frac{1 + 24\eta_{PY} G(R_{PY} s)}{R_{PY} s}\right]^{-1} \quad (1.27)$$

where η_{PY} is the volume fraction of the hard spheres and R_{PY} is a function of R, which indicates half of the lattice parameter. $G(R_{PY} s)$ is the so called "correlation function of Ornstein" of the PY equation [49,51]. The analytical expression of the averaged terms for the form factor ($\langle |F(s)|^2 \rangle$ and $|\langle F(s) \rangle|^2$) and of the structure factor employed in the PY algorithm are showed in Appendix A.4.1.

The employment of the PY model within this work is justified by the fact that this approach has proved to be suitable in characterizing both sizes and low-ordered packings of spherical pores. The most important structural and morphological parameters, which are obtained by the fitting of the scattering curves, are: the average pore radius $\langle R \rangle$, its variance σ_R, which gives information on the polydispersity of the pore size, the volume fraction η_{PY} and the radius of Percus-Yevick R_{PY}, which identifies the half of the lattice parameter, i.e. half of the pore-to-pore distance.

1.2.3 Disordered systems

An appropriate representation of the small-angle scattering for disordered systems with sharp and diffuse phase boundaries is given by the "chord length distribution" (CLD), $g(r)$. Although the concept of CLD was introduced by Porod in 1952 for the basic understanding of the small-angle scattering theory (Eqn. 1.13), it found wider application in the description of porous systems by Méring and Tchoubar at the end of the sixties [52–54]. The CLD has turned out to be an appropriate tool to characterize two-phase systems, providing important structural information. The porosity ϕ of the system, the specific surface area per volume S/V and the average pore size l_{p_1} and pore wall l_{p_2}, i.e. intrapore distance (Fig. 1.4c), are directly related to the average chord length l_p (Eq. 1.21). Moreover, as mentioned above, a value of $g(0) > 0$ reflects the angularity of the system (Eqn. 1.15). According to Ciccariello et al. [55–58] the presence of edges, characterized by an angle α and length L, indicate a value of $g(0)$ given by

$$g(0) = \frac{2\langle L \rangle}{3\pi S} \langle w(\alpha) \rangle \tag{1.28}$$

where $w(\alpha) = 1 + (\pi - \alpha)\cot\alpha$.

The chord length distribution can be described by means of the interference function $G(s)$, being its Fourier transform

$$G(s) = \mathscr{F}(g(r)), \quad \text{with}: \; G(s) \propto \left(\lim_{s \to +\infty} s^4 I(s) - s^4 I(s)\right).$$

Furthermore, $g(r)$ is the second derivative of the correlation function $\gamma(r)$ (Eqn. 1.14) and it can be directly obtained from the SAXS curve by

$$g(r) = -8 \int_0^\infty [1 - 2\pi^3 s^4 l_p I(s)] \frac{d^2 \sin z}{dz^2 z} ds; \quad z = 2\pi r s. \tag{1.29}$$

Although this evaluation approach, developed by Méring and Tchoubar, was applied in previous studies [52–54], it revealed to be improper especially in the characterization of microporosity, i.e. when the topology of the phase boundary cannot be neglected. In fact, this methodology can be strongly affected by experimental uncertainties at larger scattering vector values (Porod regime) such as data noise or smearing effects. Moreover the starting point s_{Porod} of an ideal s^{-4} behavior, i.e. ideal two phase system behavior (Eqn. 1.22), has to be set arbitrarily, which it might predetermines the outcome of the whole procedure. For these

1.2.3. Disordered systems

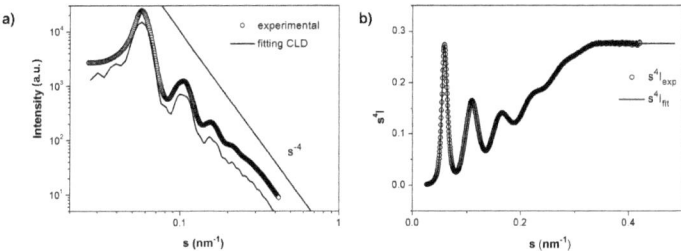

Figure 1.5: (a) Experimental scattering curve (void circle) of a bimodal silica material and the corresponding CLD fitting (bold line) following the Porod asymptote s^{-4}. (b) Porod plot of the experimental SAXS curve (void circle) and the corresponding CLD fitting (bold line).

reasons the current approach can not be completely trusted for a reliable SAXS analysis.

The procedure for the evaluation of the CLD here applied, is based on the parametrization ("fitting") of experimental scattering results by means of analytical basic functions $g_j(r/b)$ and parameters a_j, b [59]

$$g(r) = g(a_0, a_1, ..., a_N, b|r) = \sum_{j=0}^{N} a_j g_j(r/b). \qquad (1.30)$$

By using suitable basic functions, the parametrized CLD can be converted into corresponding SAXS model functions [60]

$$I_j(s) = \frac{1}{l_p \pi s^2} \int_0^\infty g_j(r) \frac{2 - 2\cos z - z \sin z}{z^2} r^2 dr. \qquad (1.31)$$

The main idea of the present approach is to fit the model function

$$I_{fit}(a_j, b_j, s) = \sum_{j=0}^{N} a_j I_j(2\pi bs) \qquad (1.32)$$

directly to the experimental SAXS curve with parameters a_j, b with the help of standard fitting and regularization methods [61–63]. The main features and advantages of this parametrization algorithm are that:
○ The CLD and the aforementioned structural parameters can be obtained directly from the parameters a_j, b;

19

1. Physical methods

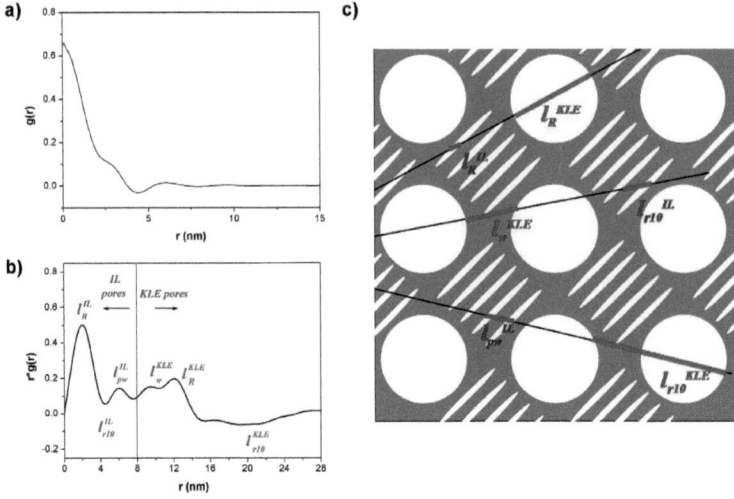

Figure 1.6: Chord length distribution $g(r)$ (a) and in the form $r^*g(r)$ (b) of a bimodal porous silica. Schematic illustration (c) of the material presenting worm-like and spherical pores. The maxima and minima indicate the distribution of the chords within the phases and penetrating the phases of the material, respectively.

○ The entire evaluation can be performed in one step;
○ The fitting approach considers the statistical data noise, performing reliable extrapolations of the data;
○ The starting point of the Porod asymptote has not defined arbitrarily as in the former method.

A typical example of CLD data treatment is presented as follows. The analysis is carried out on a porous silica sample (KLE-IL), extensively investigated during this work, bearing two different kinds of pores: worm-like (IL) of 2-3 nm in size, and spherical (KLE) of 13 nm (Fig. 1.6c). The first step of the procedure consists in the subtraction, from the experimental scattering curve, of an ordinary scattering background rising from statistical data noise and smearing effects, in order to establish an ideal Porod behavior (Fig. 1.5a). Subsequently, the parameters a_j, b have to be set in the CLD algorithm in order to verify the so-called

Porod plot ($s^4 I(s)$ vs. s, Fig. 1.5b) between the experimental and the fitted scattering data created by means of Eqn. 1.32. Finally the CLD function is created. As proof of right fitting, the simulated scattering curve has to match with the experimental one (Fig. 1.5a). A typical CLD curve is depicted in Fig 1.6a, where the relative high value of $g(0)$ reveals the presence of a defined angularity of the material. A direct representation of the phase distances in the system is given by the first momentum of the CLD, $r^*g(r)$ (Eqn. 1.13). In Fig. 1.6b the $r^*g(r)$ plot of the aforementioned material is shown. As one can see, the curve is nothing else than a superposition of the distribution of chords within the phases, visualized as maxima (e.g. l_R), and chords penetrating the phases, visualized as minima (e.g. l_{r10}). Thus, due to this merging effect, a clear interpretation of the single contributions like pore sizes (l_R), pore walls (l_w) and pore-to-pore distances (l_{r10}) becomes quite difficult, especially for complicated porous systems. For this reason, a CLD analysis should always be accompanied by complementary analyses like nitrogen sorption or TEM.

1.2.4 2D low-angle transmission SAXS

Two-dimensional (2D) low-angle transmission SAXS experiments were performed in order to study the structural organization of mesoporous thin films. Typically, these materials are highly oriented perpendicularly to the substrate but show different oriented domains on the axis in the plane of the substrate. Due to this polycrystalline organization, the pore domains are randomly oriented and the transmission SAXS pattern (incidence angle $\beta = 90$ °) will consist on rings (Fig. 1.7a). Collecting scattering patterns at low incident angles ($\beta = 5 - 10°$) the diffracted beam is transmitted through the thickness of the sample, giving information about the orientation of the stacks of the pores. Owning to the better organization in this direction, the resulting scattered pattern corresponds in spots which will identify the orientation space group of the materials (Fig. 1.7b,c). Within this work, the materials investigated by means of this technique are mesoporous films possessing spherical pores. The two most common orientation space groups for spherical systems are the face centered cubic (FCC, Fig. 1.7b) and the body centered cubic (BCC, Fig. 1.7c).

In the first stage of deposition, the pristine nanostructure of the film consisting on spheres, i.e. the block copolymer micelles, dispersed in a homogeneous matrix,

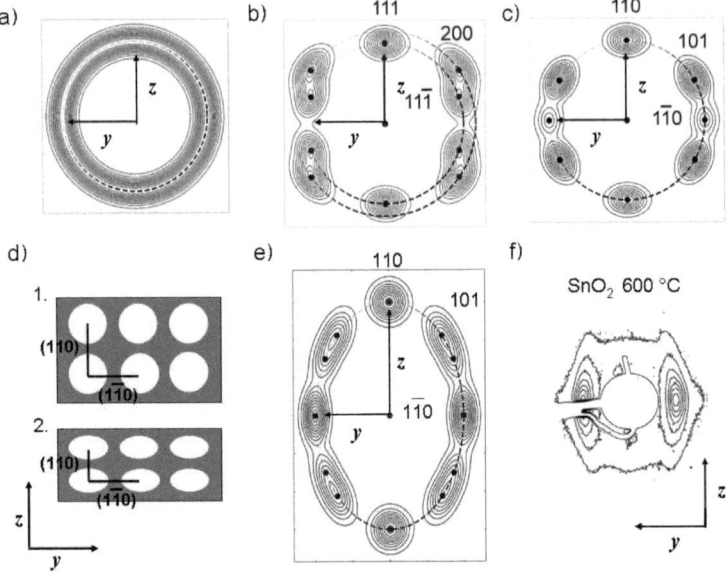

Figure 1.7: (a) 2D SAXS pattern in transmission geometry ($\beta = 90$ °) of randomly oriented pore domains; 2D SAXS of FCC (b) and BCC (c) structures of pore stacks at the pristine state; (d) representation of the shrinkage effect in a mesoporous film; 2D SAXS of a BCC structure of a shrunk film (e) and of a SnO_2 film after heating treatment at 600 °C (f) [64].

exists only within a short time interval. After some moments, the evaporation of volatile matter produces an unidirectional shrinkage of the film in the direction perpendicular to the substrate (z-axis), while in the direction parallel to the substrate (y-axis) no changes occur (Fig. 1.7d1 vs. Fig. 1.7d2). The shrinkage is further enhanced by thermal treatment and, together with a contraction of the thickness of the film, the spherical pores are transformed into spheroids of revolution. This shrinkage effect does not interferes, however, with the original structure of the material, which remains face centered or body centered if the original structures were FCC or BCC, respectively. Since the scattering patterns are expressed in reciprocal space, the contraction effect of the film will be shown as a shift to higher scattering vectors of the signal on the z-axis (cf. (110) reflection in Fig. 1.7c and Fig. 1.7e). In principle, if one would have the chance to perform scattering experiments on the pristine film, it could be possible to determine the contraction of the material through the ratio between the signals values in the z-axis before and after calcination. However, this kind of measurements can be performed only in uncommon conditions, e.g. in-situ GISAXS experiments during the self-assembly of the films at synchrotron light [65], due to the short time of life of the pristine state. Thus, an alternative way for the calculation of the shrinkage is given by the ratio between the values of the y-axis and z-axis reflections of the treated sample, the former being unchanged during calcination (cf. ($1\bar{1}0$) reflection in Fig. 1.7c and Fig. 1.7e). Usually, the contraction in metal oxide thin films is very prominent and can reach values of 60%-70% as in the case of SiO_2 [64].

In case of more complex porous systems than silica such as TiO_2, WO_3 or SnO_2, the thermal treatment can induce structural modifications, which can severely compromise the organization of the material in the z direction due to the crystallization of the metal oxide matrix. As can be seen in Fig. 1.7f, for a SnO_2 film treated at 600 °C the orientation in the z direction is lost, meaning that the structure collapses on itself and only the periodicity in the y direction, i.e. in the plane of the substrate, is preserved.

1.3 In-situ SAXS/SANS-physisorption

The in-situ SAXS/SANS-physisorption method is a combined technique, which consists in carrying out small-angle scattering measurements, using x-ray (SAXS) or neutrons (SANS) during a standard sorption experiment. The chance to simultaneously perform two different analyses is of fundamental relevance for the understanding of the physisorption behavior of gases and to precisely determine the porosity, connectivity and accessibility of porous textures. In the present work both photons and neutrons, are adopted as scattering source. As mentioned above, for neutrons holds true the same scattering theory as for x-ray. In particular, the scattering is generated through the interactions of neutrons with the atomic nuclei. In this case the scattering intensity is proportional to the square of the difference of the coherent scattering length densities (SLD), ρ_b, of the chemi-

1. Physical methods

Substance	ρ_e [10^{-5} Å^{-2}]	ρ_b [10^{-5}Å^{-2}]	Temp. [K]	Size [Å]
SiO$_2$	1.89	0.343	-	-
N$_2$	-	0.322	77.4	3.4
Kr	1.85	-	77.4 -120	4
^{36}Ar/Ar	-	0.579-0.04	77.4 - 87.3	3.7
CH$_2$Br$_2$	1.86	-	r.t.	3.5
C$_6$H$_6$/C$_6$D$_6$	-	0.118-0.5	r.t.	5
C$_5$F$_{12}$	-	0.355	r.t.	4.4/10.8

Table 1.1: Values of electron density, scattering length density, experimental temperature and atomic/molecular size for the most used contrast matcher with respect to silica. The size of the substances is expressed in terms of kinetic diameter, calculated using the approach presented in ref. [67]; in the case of C$_5$F$_{12}$ the smaller value describes the transversal direction and the larger the longitudinal one. For CH$_2$Br$_2$, C$_6$H$_6$/C$_6$D$_6$ and C$_5$F$_{12}$, r.t. stands for room temperature.

cal species (cf. Eqn. 1.20) [66]. For the sake of clarity, the following observations will be faced only in terms of scattering length density.

In section 1.2.1 it has been shown that for a two-phase system the Babinet principle of reciprocity has to be valid (Eqn. 1.19). Thus, for a porous system, the scattering intensity $I(s)$ arises from the scattering contrast $(\rho_{b_1} - \rho_{b_2})^2$ between the SLDs of the bulk material (ρ_{b_1}) and the void pores (ρ_{b_2}). If one considers, hence, to adsorb on the material a gas possessing at the liquid state $\rho_b^{ads} \approx \rho_{b_1}$, during a physisorption experiment only the void pores will contribute to SANS. In fact, since the scattering at the phase boundary between solid and adsorbed liquid (adsorbate) is negligible, the solid plus the liquid can be treated as a single phase. Consequently, once the pores are completely filled, the scattering intensity drops to zero. This effect is also called as *contrast matching* principle. In most of the previous studies on in-situ SAXS/SANS experiments the porous materials were exposed to the adsorptive (C$_6$H$_6$/C$_6$D$_6$ or H$_2$O/D$_2$O mixtures) prior to the scattering measurements, thus involving a limited accuracy for controlling the relative pressures $p/p°$ [68–70]. On the contrary, in the experiments here presented the pressure could be set with high accuracy detecting in details complicate processes of sorption in mesoporous materials like capillary condensation and pore emptying mechanisms. This chance allowed a whole understanding of both the structural characteristic of the materials and the sorption mechanism.

In principle, in-situ SAXS/SANS experiments could be realized with every material as long as a corresponding fluid is found, which, at the liquid state, satisfies the contrast matching condition. As a matter of fact, the electron density and the scattering length density are strictly dependent on the bulk density and molecular weight of the compounds. In general, it is hard to find many adequate material-fluid pairs, since in most of the cases the density of the solids overcomes significantly the one of the fluids. By chance, interesting materials like carbon and silica, being relatively light substances, do not face this problem [71–74]. For silica, since it is the main material object of these investigations within this work, the most important contrast matchers are presented in Tab. 1.1. As one can see, the chance to perform in-situ SAXS/SANS experiments with two different scattering sources allows the use of different adsorptives, from noble gases to organic molecules, of different size and at different temperatures. In this sense, the diverse experimental conditions allow to enlarge the potential outcomes from this technique. In addition, it deserves to be noted that for some probes substances like argon or benzene, a precise matching SLD is achieved by a mixture with the corresponding isotope or deuterated molecule, thus avoiding possible phenomena of mismatching, which can be observed, for instance, in the case of N_2.

1.4 Wide-angle x-ray scattering (WAXS)

The atomic structure organization of a material can be exhaustively characterized performing x-ray scattering experiments at wide angles ($5° < \theta < 60°$). In comparison with SAXS, a standard WAXS analysis is performed at higher scattering vectors and thus, due to the reciprocal space, allows the detection of objects in the atomic resolution. The general scattering theory presented in section 1.2.1 is valid also for WAXS in all its aspects, with the only variation that the investigated objects are here atoms and not particles.

When a material presents a crystalline structure, the atoms are organized in unit cells, characterized by well defined interatomic distances. These primary units are, in turn, assembled in crystallites of different size and shape. As counterpart, a non-crystalline material is defined as amorphous. In this context the scattering intensity $I(\mathbf{s})$ can be expressed in the formalism of Eqn. 1.23, being the product of an atomic form factor and lattice factor. The former member gives information about the distribution of the atoms in the unit cell, while the latter reveals the organization of the atoms in the space, i.e. the unit cell geometry.

In a diffraction pattern, the degree of crystallinity of a material can be observed by the presence of intense peaks (Bragg reflections) produced by the constructive interference between a x-ray beam and the material lattice. The analysis of the position of the Bragg peaks reveal the kind of geometry of the unit cell (cubic, orthorombic, monocline). The higher the order of the material is, the higher is the number of diffraction maxima. Every maximum is labelled with an index hkl (Miller index), which identifies a specific interatomic distance. The materials investigated through WAXS experiments within this work possess a polycrystalline structure, meaning that the crystallites are organized in randomly oriented domains. In such polycrystalline samples an approximative value of the average size of the crystallites can be obtained by the application of the Scherrer equation

$$\varphi_{cryst} = K \frac{\lambda}{\Delta \omega \cos \theta} \qquad (1.33)$$

where λ is the x-ray wavelength, $\Delta \omega$ the full width at half maximum (fwhm) of the most intensive peak and θ the position of the peak.

1.5 Electron microscopy

Electron microscopes are instruments which investigate objects in a very fine scale by means of an electron beam. This technique gives information about the topology and morphology of a material, its composition and crystallinity. The resolution d of a microscope is directly proportional to the wavelength λ of the imaging radiation. Since electrons possess wave-like properties and their wavelength is much lower than the one of the visible light, they can be used to obtain high resolution images.

Scanning electron microscopy (SEM) employs the scattering of electrons from the sample surface revealing the topology and morphology of the surface itself. In a typical SEM experiment, the electron beam is focused in a spot of ca. 1-5 nm in size. The spot is directed of the sample where the electrons interact with the sample surface, emitting secondary electrons (SE). These electrons are relatively slow and are deflected from the specimen by a weak electric field. Successively, these ones are accelerated by an amplifier and detected by an electron multiplier, depicting the sample morphology. The presence of edges within the analysed material facilitates the release of SE, thus the resulting picture appears brighter and rich in contrast. Non-conducting samples have to be coated with a conductive material (usually gold) since the electron beam charges the sample negatively.

Higher resolution can be obtained using transmission electron microscopy (TEM). In this case the electron beam passes through the entire material, requiring the preparation of very thin samples. TEM works similarly to slide projection, where a light beam transmits a slide and is affected by the picture on the slide. The transmitted beam is then projected onto a big screen forming an enlarged image. In TEM since the electrons can be either scatterd or absorbed by the sample, the instrument can operate in two different modes. In the *image mode* the scattered electrons are suppressed by a fenditure under the sample. Only the non-affected electrons pass through lenses which enlarge the beam and create a magnified image of the material. The absorption and the scattering increase with the sample thickness and the atomic number. Pores for example, do not absorb electrons and appear darker on the magnified image. The scattered electrons are used in the *diffraction mode* to achieve diffraction patterns of the observed regions. The main advantage of TEM, besides the higher resolution, is the chance to switch between these two acquisition modes. For example, the image mode reveal the presence of nanocrystal in a matrix, while the diffraction mode give information on their crystal structures. In the present work all the TEM micrographs have been acquired in image mode [75, 76].

1.6 Atomic force microscopy (AFM)

Atomic force microscopy (AFM) is a microscopy technique meant for the surface morphology characterization of a large scale of samples including ceramics, biological samples and polymers. The main advantage of AFM with respect to the other conventional microscopy techniques is the possibility to perform three-dimensional images, thus enabling, for instance, the investigations of the roughness of analysed surfaces. Furthermore AFM requires neither a vacuum environment nor any special sample preparation and it can be used both in ambient and in liquid environment.

Contact mode is the one of the more used scanning probe modes, and operates by rastering a sharp tip (usually made of Si_3N_4) across the sample. Low interatomic forces, like Van der Waals forces or capillary forces, are maintained on the cantilever thereby pushing the tip towards the sample as it rasters. The tip deflection is measured using a laser spot reflected from the top surface of the cantilever into an array of photodiodes and then converted into an analogue image of the sample surface. If the tip was scanned at a constant height, a risk would exist that the tip collides with the surface, causing damage. Hence, in most cases a feedback mechanism is employed to adjust the tip-to-sample distance to maintain a constant force between the tip and the sample. Traditionally, the sample is mounted on a piezoelectric tube that can move the sample in the z direction for maintaining a constant force, and the x-and y-directions for scanning the sample.

In the case of soft surfaces (polymer and biological samples) the sample is often destroyed or even pushed out of the field of view of the rastering tip. For this reason other kinds of imaging modes like tapping mode or lift mode, where the tip-cantilever assembly oscillates on the sample surface, can be applied [77,78].

Chapter 2

Synthesis of mesoporous metal oxides using new amphiphilic block copolymers

2.1 Introduction

Numerous natural materials present a negatively charged mineral framework bearing cavities, cages or tunnels in which small molecules like water are occluded. A family of these microporous inorganic solids is represented by the zeolites. Thanks to their own large internal surface area, zeolites find widespread applications in the field of industrial catalytic processes, as selective ion-exchange agents and sorbents. The microporosity in these materials is generated by introducing in the reaction mixture an organic molecule (generally N-based compounds), acting as *molecular template*, around which the inorganic phase is built. The large variety of applications for zeolites is however limited by the accessibility and dimension of the pores, which are restrained to the sub-nanometer scale. For this reason, during the almost two last decades an important effort has been focused on obtaining larger pore size materials.

The introduction of supramolecular assemblies, i.e. micellar aggregates, allowed the achievement of mesoporous silica compounds. These materials are characterized by ordered mesoporosity presenting a sharp size distribution between 20 and 100 Å. They find potential application in the field of catalysis, photonics, sensors and separation. Thus, since their discovery by a research group at

2. Synthesis of mesoporous metal oxides using new amphiphilic block copolymers

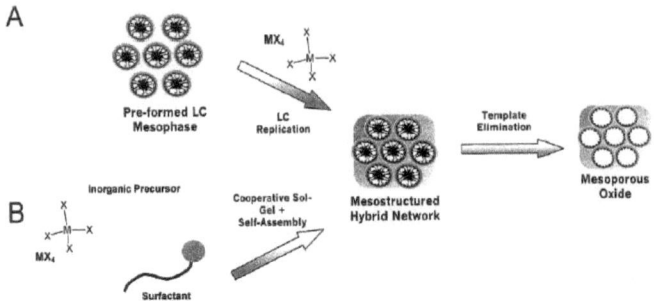

Figure 2.1: Synthetic approaches for mesostructured materials: liquid crystal templating (A) and cooperative self-assembly (B) method [79].

Mobil Oil [80, 81], a continuous effort by the scientific community was done to develop new textured inorganic or hybrid phases. The synthetic route is based through a soft chemistry approach where "cooperative self-assembly" (Fig. 2.1B) takes place in-situ between the templates, i.e. a quaternary ammonium salt, and the inorganic precursor yielding the precipitation of an organized aggregate. The formation of the mesostructured material is then governed by the electrostatic interactions between the inorganic species and the ionic template. The latter will be subsequently removed by calcination. The structuring of the surfactant molecules in micelles occurs above the critical micelle concentration (cmc), but far below the formation of a lyotropic liquid crystalline phase [82, 83]. The phase structures thus obtained can be manifold, the formation of which is a delicate combination of many factors including temperature, reaction time, concentration and pH of the mother solution.

The employment of *non-ionic surfactants* allowed the utilization of lyotropic liquid crystal phases as structure-directing media (Fig. 2.1A). In principle, the polycondensation of the water-soluble inorganic monomer (e.g. silicic acid) takes place in the ordered environment of a bulk surfactant mesophase. In this case the hydrogen bonding is considered the essential driving force for the formation of ordered inorganic materials. In comparison with the previous route, the lyotropic liquid crystal approach has numerous advantages. Being the ceramic oxide the direct cast of the liquid crystalline phase, a high control over the structure is achieved. Furthermore, the materials can be obtained as monoliths because the product is generated by solidifying a homogeneous bulk liquid crystal, instead of being precipitated from a heterogeneous solution as in the former case. The concept of liquid crystalline phase can be extended also to the utilization of amphiphilic block copolymers (ABCs) [82]. These polymers, like the molecular surfactants, form micelles if dissolved in water or proper solvents. The employment of such templates shows interesting aspects. The production of materials in monolithic form is facilitated, since the hybrid phase results more elastic and ductile. More important, the micellar aggregates own a larger size leading to larger porous texture.

The concept of casting (or nanocasting), typical for the liquid crystalline template approach, is not valid only for monoliths or powders, but also for thin films. The production of mesoporous thin films was improved and facilitated by means of the "evaporation-induced self-assembly" technique (EISA) [84]. The main methodological difference is the presence of a larger amount of volatile solvents with low viscosity in the homogeneous solution. Once the film is deposited, the evaporation of the solvent changes the system to an enriched template phase, i.e. to the situation of nanocasting. EISA technique is particularly suitable for the establishment of crystalline metal oxide mesostructures [85–88]. These materials, in comparison with amorphous silica, show interesting potential applications. Crystalline oxides are chemically, thermally and mechanically more robust. In addition, the presence of mesopores improves their intrinsic functionalities. For instance, mesoporous, crystalline TiO_2, ZnO and SnO_2 find various applications in photovoltaics [89], photocatalysis [90] or sensing [91].

Parallel to the studies on the fabrication of systems with monomodal distribution of mesopores, many efforts were also done to produce hierarchical pore morphologies [92–94]. In this context a hierarchical porous architecture is meant as a 3D arrangement of pores with sizes of different length scale, the smaller one being located in the walls between the larger pores. Hierarchical pore systems are supposed to show improvements because they combine high pore volumes and large surface area together with larger pore size [95].

In this chapter the synthesis of ordered mesoporous metal oxides by novel amphiphilic bolck copolymers is presented. The templating behaviour of these

2. Synthesis of mesoporous metal oxides using new amphiphilic block copolymers

polymers is studied through the generation of powder materials and thin films by the nanocasting technique. Furthermore, the preparation of hierarchical mesoporous systems is also faced. The following section gives a brief introduction into the recent progress in the synthesis of mesoporous metal oxides using block copolymers as template.

2.2 Current state of research: Nanocasting by liquid crystal templating of block copolymers

The self-assembly of amphiphilic block copolymers in micellar structures organized in a liquid crystalline phase revealed to be a smart way to enable nanocasting [96, 97]. Because of the size and the higher stability of the polymeric structures, it was possible to determine directly the quality and the precision of the casting process. In this context, silica revealed to be the ideal candidate to test the templating action of amphiphilic block copolymers [98]. The first works were pioneered by Göltner et al. using polystyrene-b-poly(ethylene oxide) [99] or polystyrene-b-polyvinylpyridine block copolymers [100]. Later, this work was extended by Stucky towards commercial Pluronics block copolymers [101] F127 or P123 (poly(ethylene oxide)-b-poly(propylene oxide)-b-poly(ethylene oxide)). Recently, the introduction of a semi-commercial block copolymer, namely poly(ethylene-co-butylene)-b-poly(ethylene oxide), called also as "KLE", established a superb application profile combining good chemical accessibility and high mesophase robustness [88, 102, 103]. The self-assembly of such block copolymers occurs in a variety of solvents, in particularly water, except for KLE, but also alcohols and THF, which enlarge the chemical perspectives for the inorganic framework generation.

The switch from the only partially hydrophobic poly(propylene oxide) block of the Pluronics to the more hydrophobic poly(ethylene-co-butylene) block of the KLE polymers significantly improved the templating process and admitted the establishment of more complex systems than silica, like metal transition oxides (TiO_2, CeO_2, HfO_2) [85–87, 103] and even perovskites [104] with a considerable improvement of the structural quality. The critical point in the establishment of mesoporous metal oxide systems templated with commercial Pluronics is the crystallization process. This takes place at high temperature (450-550 °C in the case of TiO_2) and is always followed by shrinkage of the mesostructure or eventually even by its collapse due to extensive growth of nanocrystals [65, 105]. Templates like KLE possess higher hydrophobic contrast, better temperature stability and promote larger pores sizes than Pluronics. These features assure a greater driving force for the self-assembly supporting a robust metal oxide gel structure [102, 103].

The use of particularly hydrophobic templates permitted also the generation of hierarchical mesoporous materials. The majority of bimodal mesoporous systems were indeed obtained by the addition of a block copolymer template and a small surfactant to the starting solution [106]. The first attempts were performed with Pluronics templates with scarce success. The reason is due to the strong affinity between the ionic surfactant and the block copolymer template which lead to the annihilation of the lyotropic phases [107]. Because of the weak hydrophobic character of the poly(propylene oxide) block, the interaction of the surfactant with F127 permits the formation of mixed micelles, followed by their dissociation increasing the concentration of molecular template [108]. The use of block copolymers like KLE instead, which presents a more hydrophobic contrast, shows different interaction with the small surfactant facilitating the formation of bimodal mesoporosity [109]. In this case the two types of micelles self-assemble to an organized "alloy" phase, with the small surfactant being located in the interstitial sites of the block copolymer lyotropic phase [110, 111].

Motivation

As one can see, the variation of the hydrophobicity contrast in amphiphilic block copolymer templates tremendously affects the generation of mesoporous frameworks. One aim of this work was to investigate the templating effect of two novel block copolymers, namely poly(isobutylene)-*b*-poly(ethylene oxide) and poly(isobutylene)-*b*-(poly(propylene oxide)-*co*-poly(ethylene oxide) labelled also as PIB6000 and PIB2300 respectively (Fig. 2.2) [1]. The templating behavior of block copolymers with extreme hydrophobic contrast composed of polyisobutylene and poly(ehtylene oxide) (PIB-PEO) was recently studied by Groenewolt et

---

[1]These block copolymer templates were granted by BASF AG in the sphere of a scientific collaboration

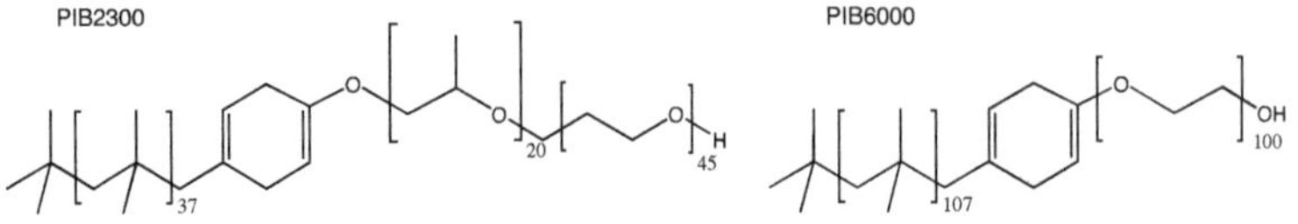

2. Synthesis of mesoporous metal oxides using new amphiphilic block copolymers

Figure 2.2: Molecular structures of the PIB-PEO block copolymers adopted in this work.

al. [112]. The casting procedure enabled the generation of robust, highly ordered, crystalline mesoporous materials with spherical pore packing of ≈ 14 nm in size. These features result interesting for potential applications in the field of materials science and technology. Thus the utilization of PIB-PEO polymers turns out to be very appealing in the preparation of functional materials. Compared to the one used by Groenewolt [112], these templates own larger chains (PIB6000) and a poly(propylene oxide-co-ethylene oxide) hydrophilic segment (PIB2300). The larger micellar size and the strong hydrophobic character suggested the chance to establish more robust mesoporous architecture with larger pore size. This can be fundamental in applications like new generation photovoltaics, i.e. Grätzel cells, where the amount of the infiltrated organic p-conductor species on the crystalline oxide is affected by its porosity and by the dimensions of the pores. In addition, in hierarchical mesostructures, thanks to their improved transport properties, larger pores could serve as transport pores, reduce back pressure and enhance diffusion.

2.3 Synthesis and characterization

In order to explore their applicability, the templating action of the novel PIB-PEO block copolymers was tested through the generation of diverse mesoporous systems, namely nanocasted powders of silica, thin films and hierarchical materials. The metal oxide replicas of the micellar aggregates were studied in terms of their dependence on template concentration.

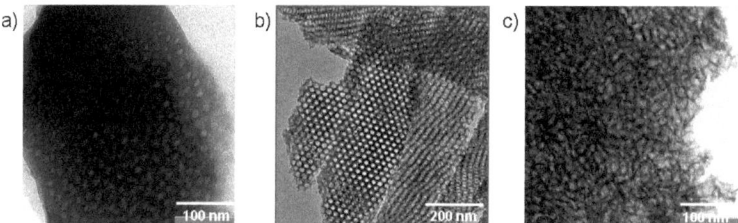

Figure 2.3: TEM images of mesoporous silica templated with 20% (a), 50% (b) and 70% (c) wt. of PIB6000.

2.3.1 Nanocasted SiO$_2$ powders

Considering the high hydrophobicity of these amphiphilic macromolecules, their dissolution by standard solvents adopted in sol-gel chemistry like water and ethanol was unsuccessful. Clear dispersions of the polymers were achieved by the utilization of a mixture 3:1 (ratio in volume fraction) ethanol/THF. The amount of THF was kept as minimized as possible, since the presence of organic solvents usually interferes with the template organization process [103]. The synthesis procedure for the generation of silica nanocasting was chosen according to the one developed by Groenewolt [112] and is exhaustively presented in Appendix A.1.1. The so prepared materials were analyzed in detail by transmission electron microscopy (TEM), small-angle x-ray scattering (SAXS) and nitrogen physisorption as function of the amount of template. The silica replicas were synthesized using different block copolymer concentrations (20% - 70% wt. with respect to the silica mass). For the sake of clarity the synthesized samples were labelled in the following way: The code P6 and P23 identifies the kind of template used, PIB6000 and PIB2300 respectively, while the number next to it represents the amount of template in units of weight percentage (e.g. P23_35).

The TEM micrographs of the corresponding most significant silica replicas for PIB6000 are shown in Fig. 2.3. For the lowest template concentration (Fig. 2.3a) spherical micelles of ca. 15 nm (Tab 2.1) with a statistically random size distribution can be observed. Increasing the amount of template, the pores get more densely packed and a growth in size is observed. At 50% wt. (Fig. 2.3b) the micelles array themselves in a face centered cubic (FCC) type micellar phase

2. Synthesis of mesoporous metal oxides using new amphiphilic block copolymers

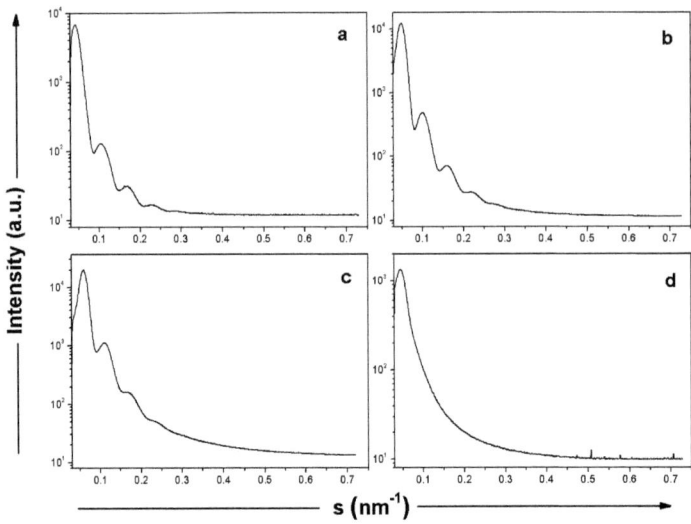

Figure 2.4: SAXS pattern of mesoporous silica templated with 20% (a), 35% (b), 50% (c) and 70% (d) wt. of PIB6000.

with high long range order, but defects of the hexagonal closed packing (HCP) type seem also to be present. In addition, the pore size enhances up to 17 nm. This order gets lost at very high fractions of template (70% wt. Fig. 2.3c). The latter situation can be considered a high porous density phase [102,113] involving deformed pores. The size of the pores has increased up to 22 nm possessing a polydisperse ellipsoidal shape. This effect at high concentration was also found, even if less prominent, in the work of Groenewolt [112] and was attributed at the border situation from a spherical cubic to a lamellar phase.

The samples were then investigated by SAXS (Fig. 2.4). It was found that the specimens prepared at low template concentration show pronounced oscillations of intensity (Fig. 2.4a, 2.4b). These oscillations are getting less pronounced with increasing the template content and disappear for the highest concentration. As described also elsewhere [102], this loss of definition is not necessarily related to the order of the system, or an increased polydispersity of the micelles, but probably to an overlap of form factor and lattice factor. A quantitative data

2.3.1. Nanocasted SiO$_2$ powders

Sample	$\otimes_{TEM}$ [nm]	$\otimes_{SAXS}$ [nm]	$\otimes_{N_2\,sorption}{}^a$ [nm]
P6_20	15	16.5	-
P6_25	16	15.2	-
P6_30	15	15.5	-
P6_35	17	17	-
P6_50	17	17	16.5
P6_70	22	-	23
P23_20	14.5	14.6	-
P23_25	14.5	14.6	-
P23_30	10	-	13
P23_35	12	-	14.5

Table 2.1: Average pore size values of PIB6000 and PIB2300 templated mesoporous silica determined by TEM, SAXS and N$_2$ sorption methods. ^{a} The pore size values were calculated by means of the non-local density function theory (NLDFT) model calculated from the adsorption branch of the physisorption isotherms by applying the kernel of metastable adsorption isotherms based on a spherical/cylindrical pore model for the system nitrogen (77.4 K)/silica.

evaluation of the scattering curves by means of the Percus-Yevick approach (see section 1.2.2) shows pores diameters according to the statistical evaluation of the electron micrographs (Tab. 2.1). The sample P6_70 could not be fitted since it possesses a too disordered and distorted structure. As also found in [102] certain deviations are present, especially in the samples with low amount of template. The reason for these deviations is unknown, but can presumably be explained by the dependence on many structural assumptions of the SAXS simulations. The shift to higher scattering vectors of the Bragg peak, increasing the template amount, evidences smaller values of the lattice parameter a and consequently more densely packed structures, as TEM images showed.

In the case of PIB2300 block copolymer, the formation of a lyotropic crystalline phase is possible up to 35% wt. of template and for higher concentration demixing throughout the silica solidification process occurs. Owing to shorter hydrophobic and hydrophilic segments of PIB2300 (Fig. 2.2), a smaller micellar size

2. Synthesis of mesoporous metal oxides using new amphiphilic block copolymers

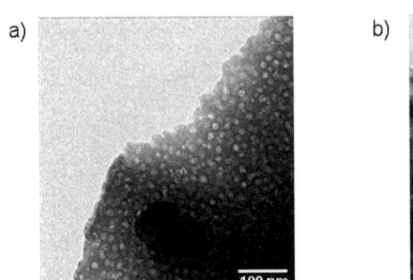
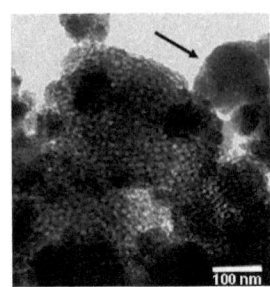

Figure 2.5: TEM images of mesoporous silica templated with 20% (a), 35% (b) wt. of PIB2300. The arrow indicates the non-templated parts of the sample.

is expected. As can be seen in Tab. 2.1, at low concentrations (20% and 25% wt.) the TEM analyses assess an average pore size of 14.5 nm. Increasing the template concentration (30% and 35% wt.), the building of a homogeneous lyotropic phase is more difficult and the system starts to phase separate. In fact in the Fig. 2.5b can be recognized templated regions and not templated ones (pointed with an arrow). The SAXS analyses in Fig. 2.6 show a cubic arrangement of the pores owning a size of ca. 15 nm, in good agreement with the TEM micrographs. Increasing the polymer content, the system gets more and more disordered (Fig. 2.6c, 2.6d), characterized by a highly polydispersed pore size distribution, thus a reasonable pore size estimation from SAXS is not possible anymore.

Aside the structural characterization, porosity studies were also performed. These investigations were possible only on samples which presented high concentrations of template, which assured sufficient connections between the pores to enable the adsorptive to flow through the matrix. The interpretation of the type IV physisorption isotherms in Fig. 2.7a, typical of mesoporous materials, points out diverse textural features of the systems. The well defined shaped hysteresis loop of P6_50 with parallel adsorption and desorption branches defines an ordered porous structure, while in P23_30 the smooth adsorption suggests a broader pore size distribution. Concerning the samples with the highest template amount, the irregular shape of the desorption branch is typical of disordered structures with undefined pore geometry [21]. These architectures possess relatively high surface areas. As assessed in Fig. 2.10c, the BET surface areas of P6_70 and P23_35

2.3.1. Nanocasted SiO₂ powders

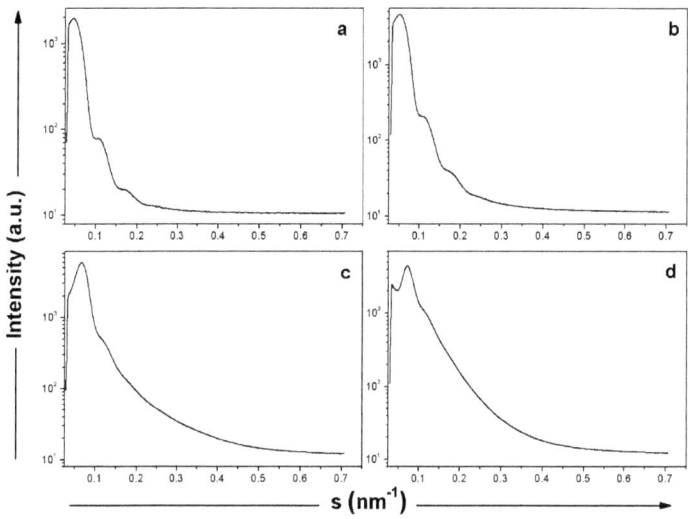

Figure 2.6: SAXS pattern of mesoporous silica templated with 20% (a), 25% (b), 30% (c) and 35% (d) wt. of PIB2300.

are almost twice of those of P6_50 and P23_30. This finding is due to both an enhancement of the template amount and an increasing of the micropores fraction, as can be seen by the larger adsorbed volume values at low relative pressure values ($0 < p/p° < 0.1$ of Fig. 2.7). This latter aspect most probably ensures a better connection between the bigger pores [106, 114], leading to higher accessible surfaces. The average pore size, obtained by means of the non-local density functional theory method (NLDFT, Tab 2.1), essentially corroborates the TEM and SAXS results. In particular, the apparently small deviation for PIB2300 samples has to be ascribed to the high polydispersity noticed also in the SAXS analyses (cf. Fig. 2.7b).

Moreover the study of the porosity of the materials with high content of template (Tab 2.2) is a matter of interest in the understanding of the templating action of the PIB-PEO block copolymers. The theoretical porous volume fraction of the materials can be calculated by the formula:

$$\phi_{Vtheo} = \frac{V_{poly}}{V_{poly} + V_{SiO_2}} \qquad (2.1)$$

39

2. Synthesis of mesoporous metal oxides using new amphiphilic block copolymers

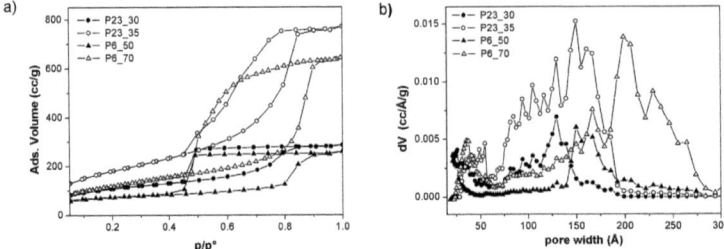

Figure 2.7: a) Nitrogen physisorption isotherms at 77 K of mesoporous silica powders with high template concentration; b) Pore size distribution of mesoporous silica powders with high template concentration calculated with the NLDFT model calculated from the adsorption branch of the physisorption isotherms by applying the kernel of metastable adsorption isotherms based on a spherical/cylindrical pore model for the system nitrogen (77.4 K)/silica.

where V_{poly} is the volume of the templating polymer and V_{SiO_2} the volume of the silica in the starting solution (see Appendix A.1.1 for more details). The maximum reachable value of ϕ_{Vtheo} for an ordered lyotropic-phase system is the one of a FCC packing of pores with extremely thin pore walls, which is $\phi_{VFCC} = 0.74$. Supposing that all the template participate to the formation of a lyotropic phase, the theoretical porous volume fraction should correspond to the experimental one ϕ_{Vexp}, which is calculated by

$$\phi_{Vexp} = \frac{V_p}{V_p + \rho_{SiO_2}^{-1}} \quad (2.2)$$

where V_p is the pore volume of the material and ρ_{SiO_2} the density of SiO_2 (ρ_{SiO_2} = 2.2 cm^3/g). The value of V_p is obtained by the N_2 sorption analysis applying the Gurvich law (Eqn. 1.2) using the volume adsorbed at $p/p° = 0.95$. From the values in Tab. 2.2 one can see that for the PIB6000 samples the ϕ_{Vexp} values are smaller than the theoretical ones. In P6_50 this finding can be explained considering that a certain amount of polymer does not act as porogen agent. On the contrary, in P6_70 the theoretical value could not be reached, due to the FCC boundary condition described above. The considerable high value $\phi_{Vexp} = 0.68$ has to be ascribed, however, to a considerable amount of disordered porosity, expressed in

Sample	V_p [cm^3/g]	V_{poly} [cm^3]	V_{SiO_2} [cm^3]	$\phi_{V theo}$	$\phi_{V exp}$
P6_50	0.385	0.072	0.036	0.66	0.45
P6_70	0.98	0.015	0.036	0.82	0.68
P23_30	0.432	0.031	0.036	0.46	0.49
P23_35	1.171	0.039	0.036	0.52	0.72

Table 2.2: Pore volume (V_p), volume of silica in the starting solution (V_{SiO_2}), theoretical and experimental porosity fractions ($\phi_{V theo}$, $\phi_{V exp}$) for the silica powders at high content of template.

terms of pore shape and pore arrangement, which is proper of this material as the former analyses assessed. In the case of the PIB2300 materials the experimental porous volume fractions are higher than the theoretical ones. Also in this case this difference can most probably due to disordered porosity, as the broad pore size distributions in Fig. 2.7b and SAXS analyses (Fig. 2.6) point out.

In the light of these results, the templating behavior for both polymers was quite different. PIB6000 casts clearly showed that the pore size increases at highest block copolymer concentration. This was also found in [102] and can be explained by the fact that the steric repulsion between different solvating chains of the same micelle becomes weaker and more densely micellar aggregates can be made [115]. In the case of PIB2300 the templating action appears quite limited. One possible explanation can be that the presence of the PPO segment smooths the hydrophobic contrast between the small PIB and PEO blocks. Furthermore, the presence of THF as solvent can disturb the micellization process especially in templates with a lower robust profile like PIB2300 [103]. On the contrary in PIB6000, thanks to the higher hydrophobic contrast, the larger PIB and PEO moieties build more robust micelles. Thus, a homogeneous lyotropic crystalline phase can be established even at extremely large amounts of template (70% wt.) without problems of demixing.

2.3.2 Hierarchical SiO$_2$ systems

The porosity of mesoporous materials is a crucial feature for their potential application. For this reason the establishment of hierarchical pore systems, being characterized by high surface areas and pore volumes, results of particular in-

2. Synthesis of mesoporous metal oxides using new amphiphilic block copolymers

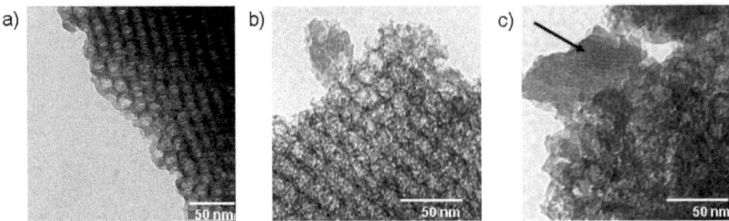

Figure 2.8: TEM images of hierarchical mesoporous silica PIB-IL 10 (a), PIB-IL 40 (b) and PIB-IL 100 (c). The arrow shows phase separated domains of IL pores

terest. Recent studies in fact showed that these systems exhibit higher activity and improved selectivity in separation processes [116, 117] as well as in showing advanced diffusion properties [118]. Sel et al. [109] demonstrated that bimodal mesoporous networks could be established by choosing block copolymers with high hydrophobic contrast, mixed together with an ionic liquid (IL), namely 1-hexadecyl-3-methylimidazolium chloride, C_{16}-mimCl [111]. In the present section the achievement of hierarchical networks by means of PIB-PEO block copolymers is investigated. Moreover it will be studied the effect of IL concentration on the structural properties of the obtained materials with special regard to the porosity. For this reason a multi-technique analysis approach (TEM, SAXS, N_2 sorption) is applied.

Showing the highest hydrophobic contrast and enabling highly ordered mesostructure, PIB6000 at the concentration of 50% wt. (P6_50) was chosen as block copolymer template system. The synthesis procedure for the bimodal architectures is mostly the same of the monomodal material, for a detailed description one can see Appendix A.1.2. Diverse samples were synthesized tuning the IL amount from 10% to 100% wt. with respect to block copolymer mass. The samples were then labelled by the abbreviation "PIB-IL" and a number next to it, indentifying the weight percentage of IL added (e.g. PIB-IL 10). The establishment of hierarchical architectures by means of PIB-PEO polymers, is shown in the TEM micrographs (Fig. 2.8). As one can see, the introduction of the cationic surfactant at a concentration of 40% wt. leads to a homogeneous distribution of cylindrical worm-like IL pores of 2-3 nm, placed in between the 17 nm ones (Fig. 2.8b). For lower concentrations (10% wt., Fig. 2.8a) the presence of small mesoporosity

2.3.2. Hierarchical SiO$_2$ systems

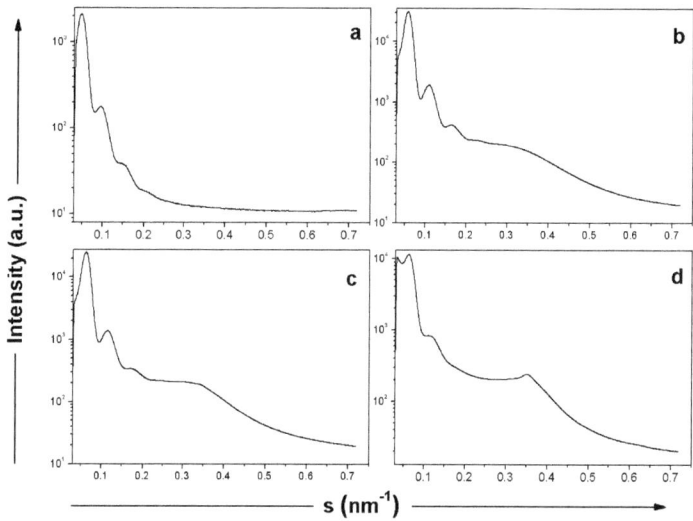

Figure 2.9: SAXS patterns of different PIB-IL silicas: PIB-IL 10 (a), PIB-IL 40 (b), PIB-IL 60 (c) and PIB-IL 100 (d).

cannot be advised. This can be explained either by a high dispersion of the pores not detectable by TEM, or because the critical IL-micelle concentration is not reached. On the other hand, if the surfactant amount exceeds, strong disorganization and deformation of the spherical pores occurs, as depicted in Fig. 2.8c. Furthermore the small IL micelles start to phase separate and the formation of cylindrical domains becomes visible (see arrow in Fig. 2.8c).

The SAXS experiments revealed that the introduction of ionic liquid in small quantities (Fig. 2.9a) does not affect the material structure and only the oscillations originated from the domains of the spherical pores (scattering vectors below $s = 0.2$ nm^{-1}) are visible. To further concentration (Fig. 2.9b, 2.9c), a broad maximum around $s = 0.3 - 0.4$ nm^{-1} is additionaly detected, originating from the mutual arrangement of the small IL mesopores. The intensification of this signal at much higher concentration of IL (Fig. 2.9d) suggests a self-organization of the small porosity, which also leads to an alteration of the spherical pore texture, in total agreement with the electron microscopy analyses.

2. Synthesis of mesoporous metal oxides using new amphiphilic block copolymers

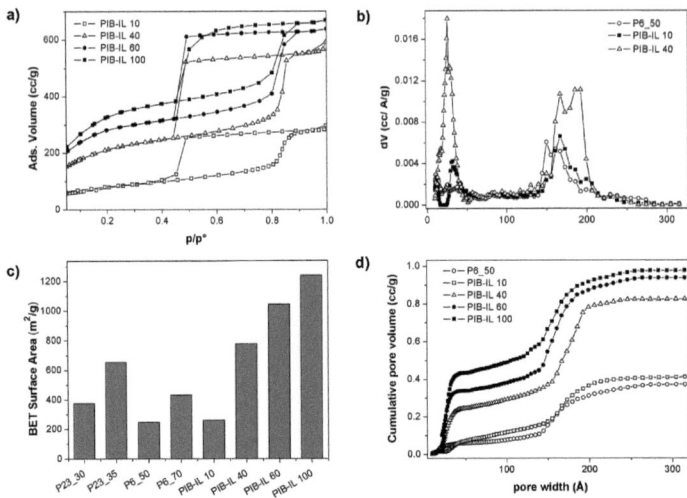

Figure 2.10: (a) Nitrogen sorption isotherms at 77 K of PIB-IL hierarchical mesoporous silicas. Pore size distributions (b) and cumulative pore volumes (d) of mono- and bimodal silicas obtained by the NLDFT model calculated from the adsorption branch of the physisorption isotherms by applying the kernel of metastable adsorption isotherms based on a spherical/cylindrical pore model for the system nitrogen (77.4 K)/silica. (c) BET surface area profiles of monomodal and bimodal silica powders. The BET plot was performed in the range $0.05 < p/p° < 0.15$ where no adsorptive condensation in the small IL pores is supposed to take place.

The nitrogen sorption analyses reveal well-defined type IV isotherms (Fig. 2.10a) possessing a hysteresis loop superimposed with the isotherm of the smaller mesopores. Moreover, the smoothly shaped desorption branch of PIB-IL 100 is typical of deformed structures, according to SAXS and TEM data. The pore size distribution analyses (Fig. 2.10b) exhibit two well defined maxima for all the bimodal materials (for the sake of clarity just PIB-IL 10 and 40 were shown), corresponding to the IL pores (2-3 nm) and the larger block copolymer ones (16-20 nm). Interestingly, it can be seen that also in PIB-IL 10 the formation of small mesopores occurs; evidently their concentration is too scarce to be detected by SAXS and TEM. Furthermore, the presence of a significant amount of microporosity can be advised. These high quality bimodal mesoporous networks own high BET surface areas and pore volumes, which rise with increasing the IL concentration (see Fig. 2.10c and 2.10d). This effect is minimal at low concentration but grows prominently by adding 40% wt. of IL. At this point the pore volume contribution of the block copolymer pores almost doubles, thanks to the connections through the ionic liquid pores which enable the complete opening of the porous network.

In conclusion, hierarchical mesoporous structures with enhanced porosity could be easily prepared. The chance to dispose of a highly hydrophobic block copolymer as PIB-PEO, building micellar aggregates with a robust profile, allows the adding of large quantity of ionic liquid (up to 60% wt.) without any loss in the textural order of the pores. The role of the IL pores is twofold. On the one hand they increase evidently the porosity of the material, on the other hand they work as connectors between the block copolymer cavities, otherwise being inaccessible, providing the whole accessibility to the porosity of the material.

Thin Films

The templating behavior of PIB-PEO polymers was also investigated by mesostructured thin films over silicon substrates through the evaporation-induced self-assembly technique [84]. Besides the model system silica, also crystalline titanium dioxide was synthesized, since such systems are interesting for applications like sensing and photovoltaics.

2.3.3 SiO$_2$ Films

A typical synthesis of mesoporous silica thin films is reported in Appendix A.1.3. The template amount was tuned from 12% wt. to 60% wt. with respect to the silica mass, and the obtained materials were investigated by means of electron microscopy (TEM, SEM), atomic force microscopy (AFM) and small-angle x-ray scattering (SAXS). The samples were labelled in the following way: A letter identifying the kind of metal oxide (S = SiO$_2$), a code for the block copolymer used (P6 = PIB6000, P23 = PIB2300) and a number meaning the template weight percentage (18 = 18% wt.), e.g. SP23_18. Scanning electron microscopy (SEM)

2. Synthesis of mesoporous metal oxides using new amphiphilic block copolymers

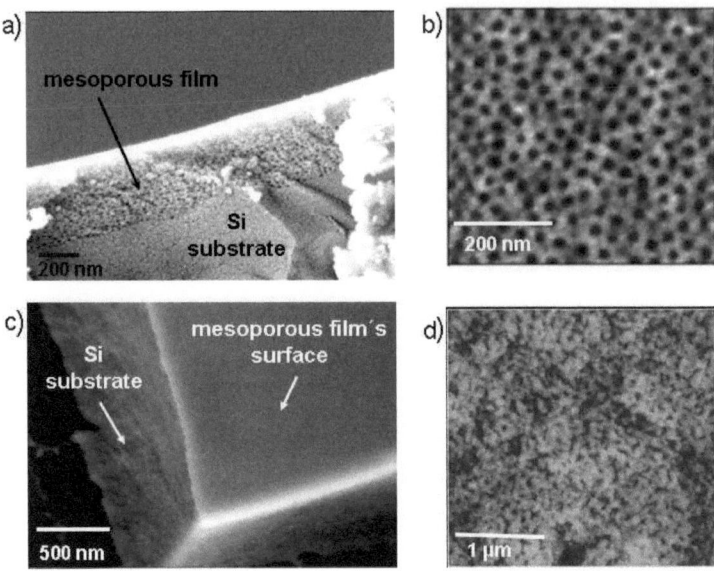

Figure 2.11: SEM micrograph of SP6_28, cross section mode, (a) and of SP6_50 (c); AFM micrograph of SP6_18 (b) and SP6_60 (d) after calcination at 550 °C.

and atomic force microscopy (AFM) were employed in the study of the film surface. As shown in Fig. 2.11, flat and crack-free coatings are generated, possessing a thickness on the order of 250 nm (Fig. 2.11a). Interestingly, no ordered mesostructure can be detected on the films' surface. The only exception in this sense is given by the sample SP6_18, in Fig. 2.11b, where spherical pores of 20 nm are regularly distributed. At lower and higher concentration than 18% wt. the absence of a visible mesostructure can be ascribed to the formation of thin bulk silica layer on the top of the well ordered pore architecture (Fig. 2.11a), as also found in a similar work [119]. For all samples, the roughness, obtained from AFM, was in the order of 1 nm only. No significant differences in the templating action of PIB6000 and PIB2300 can be found. The tuning of the block copolymer amount does not affect the film quality.

The structural characterization of the mesoporous texture was carried out by TEM and SAXS techniques. For both kinds of block copolymers TEM images

2.3.3. SiO$_2$ Films

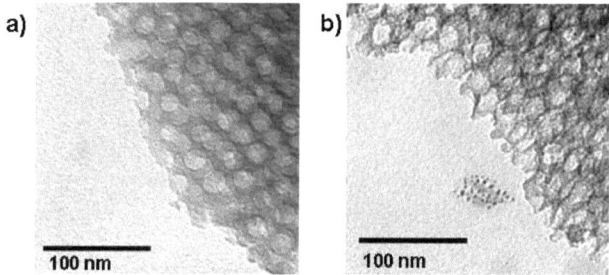

Figure 2.12: TEM images of the samples SP6_18 (a) and SP6_60 (b).

Sample	Pore size (nm)	Sample	Pore size (nm)
SP6_12	19	SP23_12	20
SP6_18	19	SP23_18	20
SP6_28	20	SP23_28	20
SP6_50	26	SP23_50	17
SP6_60	28	SP23_60	22

Table 2.3: Average pore size values of PIB6000 and PIB2300 templated mesoporous silica films determined by TEM.

(Fig. 2.12) show a homogeneous distribution of spherical pores which pack more densely increasing the amount of template. Interestingly, at the highest concentration of 60% wt. (Fig. 2.12b) the spherical pores deform into an elliptical shape similar to the corresponding powder materials. Regarding the pore dimensions, it can be clearly seen in Tab. 2.3 that for both block copolymers the pore size is of 20 nm up to 28% wt. of template, according to the AFM results. Increasing the template ratio, PIB6000 systems increase their pore size up to 28 nm while in the PIB2300 ones it oscillates between 17 nm and 22 nm. Interestingly the film systems are characterized by larger pore sizes compared to the powders. This finding can be explained considering the solvent removal during the gelation process. In the case of films, the dip-coating process permits a faster solvent evaporation than in the pot-synthesis of the powders, thus allowing the formation of micelles possessing larger size.

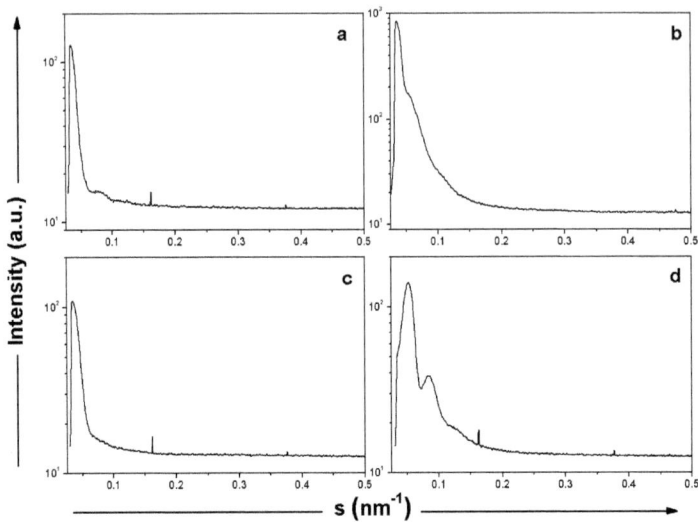

Figure 2.13: SAXS pattern of SP6_12 (a), SP6_60 (b), SP23_12 (c) and SP23_60 (d).

The porous silica films were then further studied by means of SAXS. At low template concentration (Fig. 2.13a, 2.13c) the patterns present a Bragg peak at very low scattering vectors ($s = 0.045$ nm^{-1}) and weak form factor contributions, identifying a primarily organized cubic lattice of spherical pores. Increasing the template ratio, the porous texture, becoming more dense, denotes more defined scattering patterns with form factor oscillations of the spherical pores which can be easily distinguished. Interestingly at 60% wt. of template (Fig. 2.13b) the PIB6000 system owns a quite smeared pattern typical of highly polydispersed architectures. These findings were verified by a quantitative interpretation of the scattering curves (Fig. 2.14b). The average pore diameters (white spots) were obtained by the application of the Percus-Yevick model, described exhaustively in section 1.2.2. The fitting of the experimental curves is shown in Fig. 2.14a. The lattice parameter a of the cubic structure, namely the pore-to-pore distance (black spots), was obtained by the reciprocal value of the scattering vector of the Bragg peak maximum. As one can see for both block copolymers up to 28% wt. of template a progressive decreasing of the pore distances occurs and the

2.3.3. SiO₂ Films

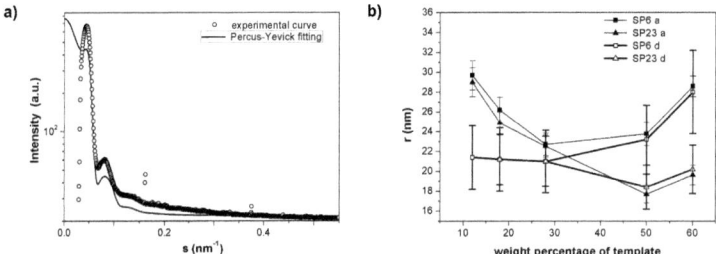

Figure 2.14: (a) Structural model fitting (bold line) and experimental curve (void circles) of the sample SP6_28. (b) Average pore diameter of SiO_2 films determined by the Percus-Yevick model (see section 1.2.2)(SP6d, SP23d white spots) and lattice parameter (SP6a, SP23a black spots) values with respect to the template amount. The error bars in the case of the pore size (15%) are are given by the polydispersity values,corresponding to the variance σ_R, of the pore size, calculated through the Percus-Yevick approach. In the case of the pore to pore distances the error bars are set at a standard value of 5% (uncertainty in data-reading).

pore size is almost kept constant at 21 nm. The systems react quite differently to further increasing of the template amount. As observed in the powder materials, the micellization of PIB6000 chains leads to larger pores while in the case of PIB2300 the pore size slightly decreases. This can be again explained by the fact that at high concentrations in PIB6000 the steric repulsion of the different chains is blunted, allowing the formation of larger micellar structures [120]. In addition, the pore walls thin down and the lattice parameter becomes comparable to the pore size. It is noteworthy to point out the reliability of the Percus-Yevick model for the analysis of the SAXS curves showing good agreement with the TEM micrographs.

In order to further study the structural changes in the materials with respect to the pore arrangements in the space, low-angle 2D SAXS experiments were carried out (see section 1.2.4 for theoretical explanations). For all films the measurements performed at $\beta = 90°$ (Fig. 2.15a) are characterized by isotropic diffraction rings, typical of random orientation of different mesostructured domains with respect one to another in the direction parallel to the film substrate.

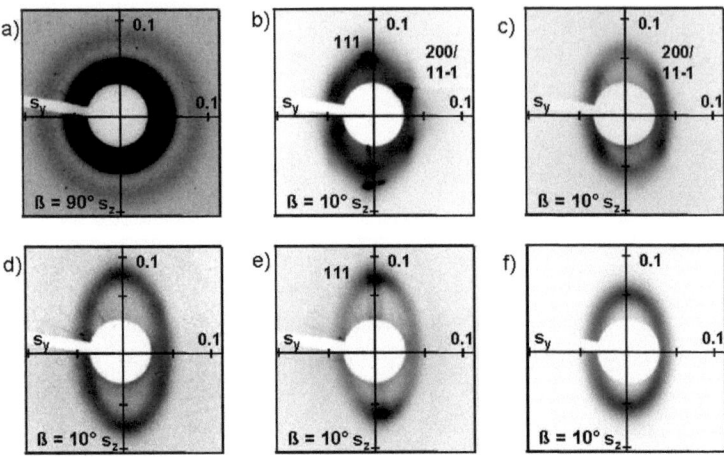

Figure 2.15: 2D SAXS patterns of silica films templated with different amount of PIB6000 block copolymer, recorded at different angles of incidence β. 28% wt. (SP6_28) $\beta = 90°$ (a), 12% wt. (b, SP6_12), 18% wt. (c, SP6_18), 28% wt. (d, SP6_28), 50% wt. (e, SP6_50), 60% wt. (f, SP6_60). For b-f $\beta = 10 \pm 1°$. Scattering vector s components in units [nm^{-1}]. In Fig. b the high-right corner was slightly graphically manipulated due to defects during the pattern acquisition.

At low angle of incidence, i.e. $\beta = 10°$ (Fig. 2.15b - 2.15f), the structure in the direction perpendicular to the substrate is studied. The patterns of the materials own features of a distorted FCC packing of spherical mesopores [64]. For the sake of clarity only the analyses of PIB6000 templated films are shown, since the PIB2300 ones present no significant differences. The low-angle measurements, compared to the ideal circular shape of the pristine structure of the film, possess an elliptical shape related to a certain degree of contraction of the mesoporous structure due to aging and calcination. Furthermore, the faded bands in the 2D SAXS images are attributed to stacking faults of the pores [64, 119]. It can be seen that the variation in the template amount clearly modifies the structural organization of the materials. At 12% and 28% wt. (Fig. 2.15b, 2.15d) the (111) and the (200)/(11-1) reflections indicate that at low template amount the system is organized in a FCC structure, even if the marked streaks evidence high density of stacking defects. These irregularities decrease in the 18% wt. sample (Fig. 2.15c) in which, however, the (111) refections are absent. SP6_18 possesses a more ordered structure which leads to a narrower (111) reflection. Thus, most probably, the incidence angle used here ($\beta = 10°$) does not represent the minimum angle for which all the lattice signals are visible [121, 122] as for the other cases. An analogous situation is encountered at 50% and 60% wt. of template (Fig. 2.15e, 2.15f) where the missing reflections are the (200)/(11-1). This case is very similar to the one presented by Sel et al. [119] where this effect was obtained by the introduction of a second template, responsible for a partial deformation of the FCC lattice. The 2D hexagonal layers of spherical mesopores, constituting the cubic stacks, does not exhibit translational correlation to the layers above and below, but a turbostratic order. In our case this finding can be regarded to the high amount of template in the material, which hinders the correct packing of spheres. In addition, the intensive streaks in SP6_60 pattern can be related to structural deformation proper of this material as 1D SAXS and TEM already assessed. It is noteworthy to point out that at high concentrations (50% and 60% wt.), even if the two templates generate pores with diverse sizes, the materials own the same 2D SAXS pattern, thus showing that the micelles packing depends on the template ratio and not on morphological parameters.

As discussed above, the aging and calcination processes lead to shrinkage of the silica mesoporous structure. Thus, depending on the elliptical form of the 2D SAXS pattern, the degree of structural contraction as function of the amount of template can be studied. The contraction ratio of the mesoporous structure is obtained as described in section 1.2.4. The obtained results (Fig. 2.16a) are typical of silica films, as also found by Ruland [64], and evidence a pore shape transition from spheres to prolates (Fig. 2.16b) [64]. As one can see there is a direct correspondence between the amount of template and the film shrinkage. At low concentrations PIB6000 systems shrink more, but upon further increasing of concentration both systems own similar profiles. The decrease in contraction after 28% wt. is most probably due to a dispensable amount of polymer which does not generate lyotropic crystalline phase and phase separates, as also found for the powder systems.

In the case of the powders materials the chance to dispose of hardly robust and stable lyotropic phase gives the opportunity to establish bimodal materials

2. Synthesis of mesoporous metal oxides using new amphiphilic block copolymers

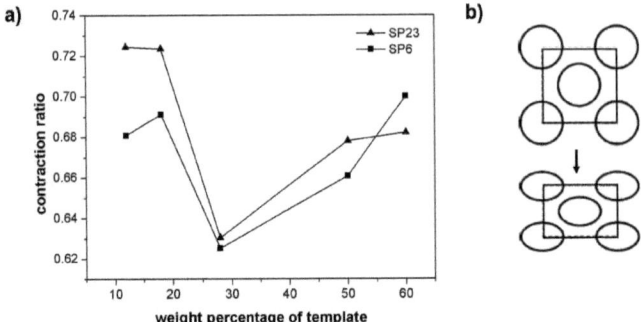

Figure 2.16: (a) Matrix contraction values for PIB6000 and PIB2300 templated mesoporous silica obtained by the 2D-SAXS pattern. (b) Scheme of the contraction of the pores in in a mesoporous film [64].

by adding an ionic template. The feasibility of bimodal silica thin films by using highly hydrophobic block copolymers was recently proved, showing a strong enhancement of the functional properties of the material [119]. PIB-PEO bimodal systems were then prepared with the same philosophy of the powders, by adding the 20% wt. of ionic liquid (C_{16}mimCl, IL) with respect to block copolymer weight (see Appendix A.1.3). As hosting monomodal system SP6_18 was chosen. The successful generation of bimodal porous films, also in the case of PIB-PEO polymers, was proved by TEM analysis (Fig. 2.17c). As one can see, besides the 20 nm spherical PIB-PEO mesopores a homogeneous dispersion of 2-3 nm worm-like IL pores is visible. The addition of a second template does not disturb the structural features of the material which results with crack-free morphology, flat surface and unaltered mesopores conformation, as SEM in Fig. 2.17a shows. The latter feature is clearly evidenced by SAXS experiments (Fig. 2.17b), showing a isotropic diffraction ring typical of the monomodal systems. Interestingly, no scattering signals attributable to the IL pores are present, since their concentration is too low to be detected by this technique.

2.3.3. SiO$_2$ Films

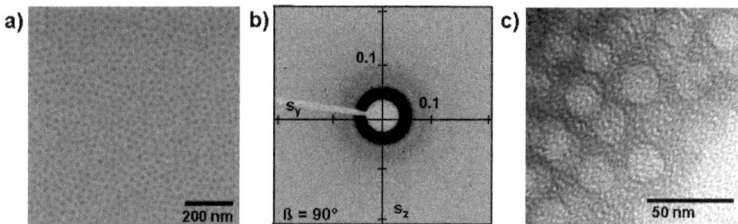

Figure 2.17: Bimodal mesoporous silica film analysed by SEM (a), SAXS (b) and TEM (c). Scattering vectors s components in units [nm^{-1}].

In this section was demonstrated how high quality mesoporous thin films could be prepared by means of PIB-PEO block copolymers. The materials present a very smooth surface, absence of cracks and an ordered organization of the porous structures. Both polymers lead to templated pore structures even at particularly high concentration (up to 60% wt.), featured by distorted face centered cubic lattices. The enhancement of the polymer amount affects, on the other side, the topology of the materials. The PIB6000 templated films are characterized by an increase of the pore size, while for PIB2300 systems it slightly decreases. This difference is most probably due to the diverse hydrophobicity of the two macromolecules. Finally, it has been shown how PIB-PEO polymers can be used to obtained well structured hierarchical porous films.

Interestingly, both for powders and thin film materials it has been shown how the templating action of these novel PIB-PEO polymers allows only the generation of spherical micelles. This finding can be considered quite singular, since other block copolymers, like Pluronics, change the morphology of the lyotropic phases by increasing of their concentration in solution. The typical morphology transition is spherical → cylindrical → lamellar [123,124]. The self-assembly of the amphiphilic block copolymers can be described in terms of the surfactant *packing parameter* [83, 125]. In this case the resulting structure of the amphiphile is parametrized as $v/\bar{a}l$, where $\bar{a}$ is the area per hydrophilic group, l is almost the fully extended length of the hydrocarbon chain, and v is the volume occupied by the surfactant. The packing parameter gives a critical condition below which

53

certain micellar shapes are stable, i.e

$$\frac{v}{\bar{a}l} < \begin{cases} 1/3 \text{ for spherical micelles} \\ 1/2 \text{ for cylindrical micelles} \\ 1 \text{ for lamellar micelles} \end{cases}$$

For a given block copolymer the values of l and $\bar{a}$ can be considered independent from the total amphiphile concentration [125]. The only parameter which changes is the volume occupied by the surfactant v. Considering that PIB6000 and PIB2300 have a high hydrophobic contrast, the poly(isobutylene) block will arrange itself in order to occupy the lowest possible volume, trying to be segregated, devoid of any solvent. Most probably the hydrophobicity of these polymers is so high that the value of v is kept remarkably small even at high concentration of surfactant, thus permitting only spherical micelles' aggregates.

2.3.4 TiO$_2$ Films

After having shown the general templating behavior in the establishment of mesoporous silica films, the PIB-PEO polymers are here used in the generation of most interesting systems like mesoporous crystalline titanium dioxide films.

As in the case of silica the templating behavior was studied with respect to the variation of template amount (28%, 40% and 50% wt.). The synthesized materials were then treated in muffle oven at 550 °C and 600 °C to get rid of the organic part and enable the crystallization of the matrix. The films were studied through microscopy techniques (SEM, TEM, AFM) and scattering methods at small and wide angle (SAXS, WAXS). The samples were labelled in the following way: A letter identifying the kind of metal oxide (T = TiO$_2$), a code for the block copolymer used (P6 = PIB6000, P23 = PIB2300), a number meaning the template weight percentage (28 = 28% wt.) and the calcination temperature (550 = 550 °C) e.g. TP23_28_550. The comprehensive synthetic route is shown in Appendix A.1.4.

The surface of the materials after calcination was analyzed by means of AFM and SEM methods. All samples present a thickness between 250 - 300 nm, a crack-free top surface with a homogeneous distribution of opened spherical pores (Fig. 2.18a - 2.18c). Surface profile analyses further assessed a roughness in the order of 2 nm. By increasing the amount of templating polymer, interesting differences

2.3.4. TiO$_2$ Films

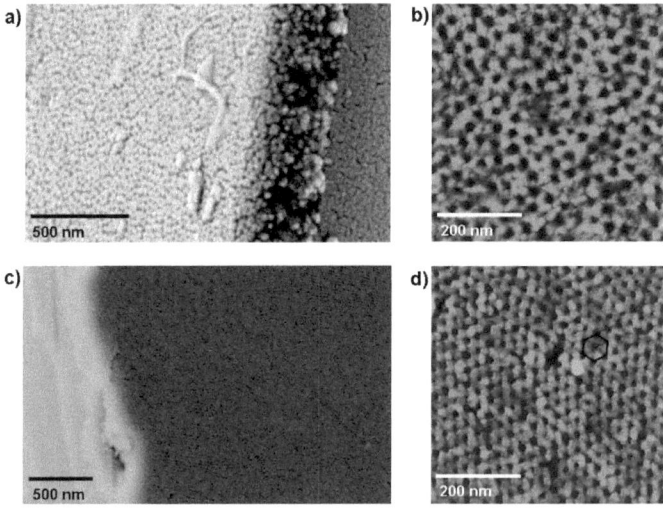

Figure 2.18: SEM and AFM analyses of different TiO$_2$ materials. Cross section SEM of TP6_28_550 (a); AFM of TP23_28_550 (b); SEM of TP6_50_600 (c); AFM of TP23_50_550 (d).

of the morphological features can be faced (Fig. 2.18b and 2.18d). In fact, the samples with higher amount of template (50%) show a better arrangement of the pores in the space, characterized by hexagonal planes typical of a cubic pore packing. Parallel to this, AFM and SEM analyses also evidenced a contraction of the pore size from 20-25 nm to 16-18 nm for the systems owning the 28% and the 50% wt. of template respectively. The different heat treatments (550 °C, 600 °C) do not affect significantly the pore size and the surface quality of the materials.

The structural information collected by AFM and SEM were corroborated also by TEM analyses, which exhibit pore size of 20-22 nm and 16-18 nm for the 28% wt. and 50% wt. templated systems. In addition, the systems owning the larger amount of template show a better organized porous structure with randomly oriented pores domains (Fig. 2.19b), according to AFM. TEM analyses revealed to be an useful tool also for the study of the microstructural changes during calcination. When the crystallization of the TiO$_2$ matrix is reached the physical cohesion of the amorphous framework is broken through the formation

55

2. Synthesis of mesoporous metal oxides using new amphiphilic block copolymers

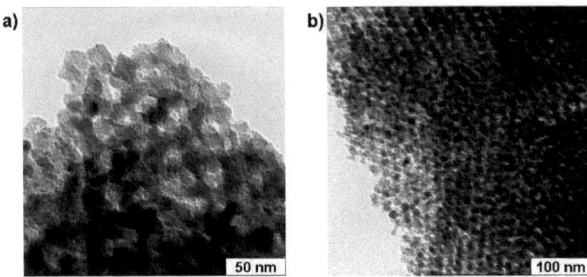

Figure 2.19: TEM micrographs of TP6_40_600 (a) and TP6_50_600 (b).

of crystalline nanoparticles (Fig. 2.19a darker regions) with the effect that the mesoporous structure gets ruined and the pores tend to merge [65]. The crystallite size of the systems resulted to be independent from the calcination temperatures (550 °C and 600 °C) and polymer amount and was estimated between 10-13 nm, in good agreement with the similar TiO_2-KLE system [103]. The achievement of a high crystallinity and at the same time the mantainance of a well ordered mesostructure is strictly dependent on the heat treatment procedure. Thus, the films underwent a pretreatment at 300 °C and a further calcination for short period of time at a temperature far above T = 450 °C, which is the temperature of onset of TiO_2 crystallization into the anatase phase [65].

The crystallinity of the metal oxide structures was deeply investigated through WAXS experiments. All the samples after heating treatments at 550 and 600 °C present the distinctive reflections of the anatase phase (Fig. 2.20a). Although the data do not allow a precise quantification of the crystallinity, the amount of amorphous TiO_2 was estimated to be in the order of only several percent in volume fraction at most, as also shown in ref. [103] for similar systems. The crystalline particle average dimension was determined by the Scherrer equation (see section 1.4) calculating the full width at half maximum (fwhm = Δs) of the (101) anatase peak. Its evolution was studied for both temperatures of calcination (550 °C and 600 °C) and shown in Fig. 2.21. As one can see, for the samples at 28% and 40% wt. of template the nanoparticles' dimension is between 12-14 nm for both temperatures, in agreement with the TEM measurements. On the other hand, the samples prepared at 50% wt. of template are characterized by crystallites of

2.3.4. TiO$_2$ Films

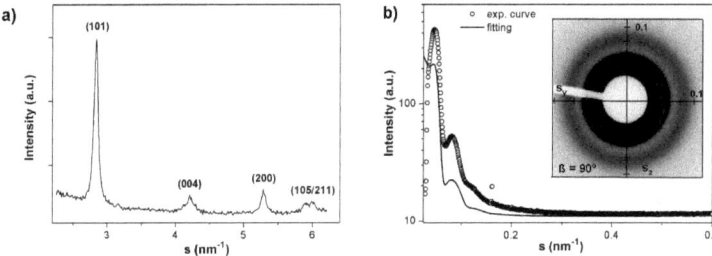

Figure 2.20: (a) WAXS pattern of TP6_40_550 revealing the typical anatase reflections. (b) SAXS pattern (experimental curve and theoretical fitting) and its 2D projection (right panel) of TP6_50_550. Scattering vectors s components in units [nm^{-1}].

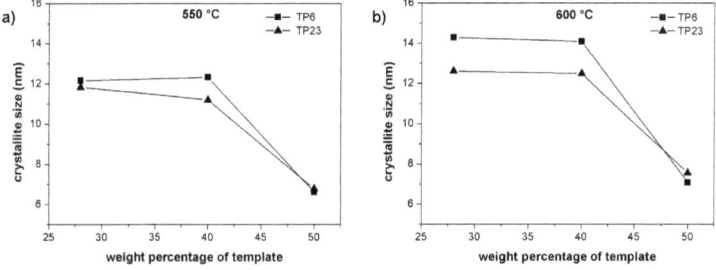

Figure 2.21: Mesoporous titania crystalline particle size variation as function of the template amount at 550 °C (a) and 600 °C (b).

only 7 nm in size. The explanation to this effect can be twofold. From one side these systems present a much more densely packed mesostructure with thinner walls (see SAXS analysis below), limiting the mass transfer during the calcination process. On the other side this highly dense packing induces the PEO chains of the micelles to deeply penetrate in the metal oxide matrix, thus hindering the direct contact between the crystallites and inhibiting their growth. The apparent incompatibility of this outcome with the previous TEM analysis, which identified crystallites of 10-13 nm in size, can be motivated by the rare chance to detect

57

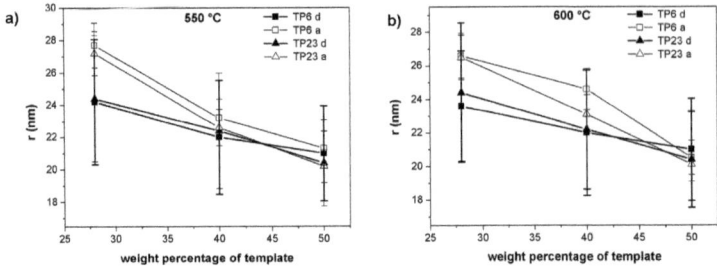

Figure 2.22: Average pore size of the TiO$_2$ films determined by the Percu-Yevick model (see section 1.2.2) (TP6d, TP23d; black spots) and lattice parameter (TP6a, TP23a; white spots) values with respect to the template amount at 550 °C (a) and 600 °C (b). The error bars in the case of the pore size (15%) are are given by the polydispersity values, corresponding to the variance σ_R of the pore size, calculated through the Percus-Yevick approach. In the case of the pore-to-pore distances the error bars are set at a standard value of 5% (uncertainty in data-reading).

single crystallites in the micrographs. In addition, a slight growth of the particles between 550 °C and 600 °C can be observed. This means that the system did not reach the complete crystallization state yet, which is known to be at T ≈ 700 °C for such systems [65].

In order to properly comprehend the mesostructure evolution as function of the amount of polymer and temperature variations, SAXS experiments were performed. All the samples are characterized by an intense Bragg peak and form factor oscillations typical of a cubic packing of spherical pores (Fig. 2.20b). The 2D SAXS measurements performed at $\beta = 90°$ provide isotropic diffraction rings typical of a random orientation of different mesostructured domains with respect one to the other in the direction parallel to the substrate. The SAXS curves were then quantitatively analysed in order to determine structural parameters like pore size and lattice parameter, namely the interpore distance, as function of polymer amount and crystallization temperature. As in the case of SiO$_2$, the pore size was obtained by the application of the Percus-Yevick (PY) model (see. section 1.2.2), while the lattice parameter from the reciprocal value of scattering vector of the Bragg peak reflection. As one can see, for both crystallization tempera-

2.3.4. TiO$_2$ Films

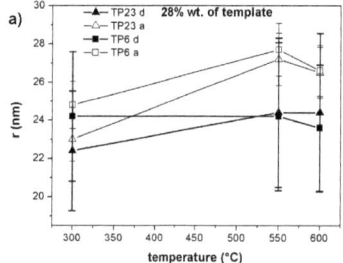
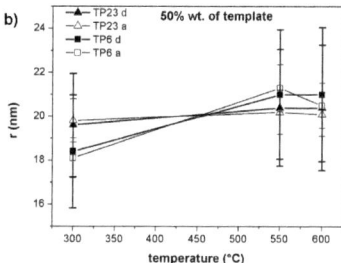

Figure 2.23: Average pore size of the TiO$_2$ films determined by the Percu-Yevick model (see section 1.2.2)(TP6d, TP23d; black spots) and lattice parameter (TP6a, TP23a; white spots) values with respect to the template amount at 28% wt. (a) and 50% wt. (b) of template. The error bars in the case of the pore size (15%) are are given by the polydispersity values, corresponding to the variance σ_R of the pore size, calculated through the Percus-Yevick approach. In the case of the pore-to-pore distances the error bars are set at a standard value of 5% (uncertainty in data-reading).

tures, increasing the polymer content, a decreasing of the pore size from 24 nm to 21 nm is observed (Fig. 2.22 black spots). These values are slightly higher than the ones observed for the other analysis techniques. This effect can be explained considering the high polydispersity in size of the pores. In fact, the scattering intensity being proportional to the volume of the scattering bodies, larger pores will contribute more than the smaller ones on the resulting scattering pattern. Interestingly, it can also be seen that the lattice parameter of the cubic packing of pores decreases as well (Fig. 2.22 white spots). This can be read as result of more densely packed structures, as the AFM images showed (cf. Fig. 2.18b vs. 2.18d). At higher concentrations the lattice parameter values tend to coincide with the pore sizes, apparently loosing physical meaning. Nevertheless, being the pore sizes slightly overestimated due to polydispersity, these results can still be considered reliable. Structural studies were carried out also as function of the materials' calcination temperature. It can be noted, that the structural evolution of the lattice parameter for low content (Fig. 2.23a) and high content (Fig. 2.23b) template materials is considerably different. At the amorphous state (300 °C) the values are close to the pore sizes, evidencing thin pore walls. As the temperature increases, the lattice factor value increases with respect to the pore size for the 28% wt. templated materials but not for the 50% wt. ones. Most probably, the crystallization process for the former films, being followed by mass transfer, lead to thicker pore walls and consequently to larger lattice parameters. For the latter systems instead, as also WAXS analyses pointed out, being the mesoporous structure more densely packed, the mass transfer is discouraged and the pore walls can not grow more. Looking at the pore size variation for both kind of polymers, taking into account the high degree of polydispersity which affects the values, no meaningful changes occur during the heating treatments. Thus, the pore size can be considered calcination-independent, as microscopy analyses already assessed.

Also in the case of SiO_2, 2D SAXS analyses were carried out. The evolution for low content template films (28% wt.) with respect to the temperature variation is presented in Fig. 2.24a,b,c. At 300 °C (Fig. 2.24a) the off-plane (1-10) and the in-plane (110) reflections, typical of a body centered cubic (BCC) structure, are observed. The absence of the (101) reflections (cf. section 1.2.4) and the presence of heavy streaks are attributed to stacking faults in the BCC lattice [64, 65]. The vertical extension of the pattern can be read as the typical unidirectional mesostructural shrinkage in the in-plane (110) direction, commonly observed during aging/calcinations of thin films [64, 126]. To further temperatures (550 °C Fig. 2.24b and 600 °C Fig. 2.24c), the loss of the in-plane (110) reflection and the steeply decrease of intensity along the z axis is due to highly disordered structures and defines the limit of the three dimensional cohesion. This means that at this temperatures the structure looses its integrity and the system undergoes pore fusion [65, 103, 127]. On the other hand, no changes can be observed in the y axis, suggesting that the periodicity is entirely retained in the (1-10) direction up to the maximal available temperature of 600 °C. These findings are corroborated by the TEM and AFM analyses where, to the detriment of a loss of integrity and the following merging of the pores (Fig 2.19a), the system still keeps its two dimensional order (Fig. 2.18b).

The study of the mesopores' structural organization for the films possessing high content of template (50% wt.), as function of the calcination temperature, is shown in Fig. 2.24d,e,f. As for the 28% wt. templated materials, after the pretreatmet at 300 °C the films possess the typical features of the contracted BCC structure (Fig. 2.24d). Overtaking the anatase crystallization temperature (550

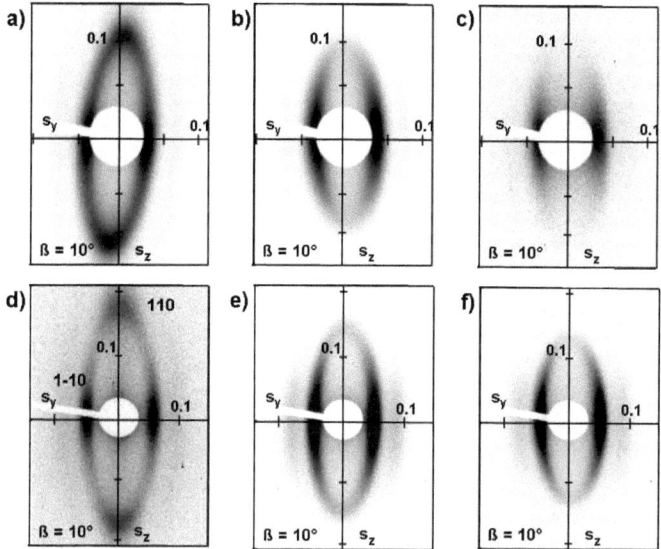

Figure 2.24: 2D SAXS patterns at $\beta = 10°$ of titania films. TP6_28_300, (a), TP6_28_550 (b) TP6_28_600 (c). TP23_50_300 (d), TP23_50_550 (e) and TP23_50_600 (f). Scattering vector s components in units [nm^{-1}].

°C Fig. 2.24e and 600 °C Fig. 2.24f), the system looses the (110) in-plane orientation, but still conserves a faint ellipsoidal circle. This corresponds to domains owning the same structure, but random orientation in the direction perpendicular to the substrate. Thus, as seen in the corresponding AFM and TEM micrographs, these materials, compared to the corresponding 28% wt. templated ones, are characterized by a better mesostructural order, which implies a more robust lattice. This phenomenon may be explained by the limited migration of matter through diffusive sintering, bringing to smaller size crystallites, as the WAXS analyses proved, which avoids the entire collapse of the mesostructure.

In this section it was shown that it is possible to achieve good quality crystalline metal oxide films through the templating action of PIB-PEO polymers. The different templating action of the two macromolecules, pointed out by the silica materials, was here not verified. The most possible explanation to this finding is that the assembly patterns of the hydrated titanic acid species are much different compared to those of the silicic acid. The latter indeed creates assembly patterns very similar to the ones found in water [128], enabling to detect templating differences. Furthermore, the heavy structural modifications which occur to the TiO_2 matrix during the heating treatment procedures can contribute also to smooth out potential peculiarities. Nevertheless, it was possible to generate lyotropic liquid crystalline phases with high amount of block copolymer, leading to the formation of materials with improved structural features compared to the low content ones.

2.4 Mesoporous TiO_2 thin films as photovoltaic devices

2.4.1 Current state of research

Solar energy is one of the most promising future energy resources. The direct conversion of sunlight into electric power by solar cells is of particular interest because it has many advantages (e.g. absence of green-house gases) with respect to the most presently used electrical power generation methods. At present, p-n junction solar cells are the most produced devices with the highest efficient light-to-power conversion. Nevertheless, new generation photovoltaic devices, namely dye sensitized solar cells (DSSC), appear to have significant potential as a low-cost alternative to conventional p-n junction solar cells [129]. The DSSC device is based on a 10-μm-thick, optically transparent porous film of titanium dioxide particles of few nanometers in size, coated with a monolayer of a charge transfer dye (usually a Ru dye) to sensitize the film for light harvesting [130]. The semiconductor is deposited onto a transparent conductive oxide (TCO) electrode, through which the cell is illuminated. The TiO_2 pores are filled with a redox electrolyte (I^-/I_3^-) which acts as conductor and is electrically connected to a platinum counterelectrode (Fig. 2.25a). Dye sensitized cells differ form the conventional p-n junction devices in that they separate the function of light absorption from charge carrier transport. The role of the ruthenium complex is the same of the chlorophyll in the green leaf: It must absorb the incident sunlight and exploit the light energy to induce an electron transfer reaction. The TiO_2 film,

2.4.1. Current state of research

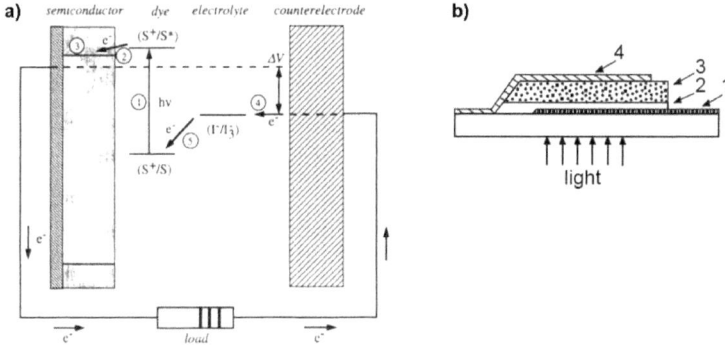

Figure 2.25: (a) Schematic representation of the principle of a dye sensitized solar cell. The cell voltage observed under illumination corresponds to ΔV. S sensitizer; S* electronically excited sensitizer; S^+ oxidize sensitizer. (b) Structure of a solid state dye sensitized solar cell: 1. conducting TCO glass; 2. compact TiO_2 layer; 3. dye sensitized heterojunction; 4. gold electrode [132].

apart from acting as support for the sensitizer, also functions as electron acceptor and electronic conductor. The electrons injected from the sensitizer into the TiO_2 conduction band travel across the nanocrystalline film to the TCO support serving as a current collector. To complete the circuit, the dye must be regenerated by electron transfer from the I^-/I_3^- redox species in solution, which is then reduced at the counterelectrode [89].

Even if DSSC devices are commercially available, market expansion did not take place because of technological issues induced by the use of highly corrosive liquid electrolytes (cell cealing, handling and maintenance). Therefore, solid state dye sensitized solar cells have been developed [131, 132] substituting the liquid electrolyte with hole transfer materials like ionically conductive gels, inorganic p-type conductors or organic hole-conductive polymers [133–135]. The scheme of this new kind of solar cell device is presented in Fig. 2.25b. The working electrode consists of a TCO substrate onto wich a compact TiO_2 layer was deposited through spray pyrolysis or sputtering. This procedure avoids the direct contact between the electrode and the hole transfer material (HTM) which would short-circuit the cell. After the deposition of a several μm-thick TiO_2 mesoporous

2. Synthesis of mesoporous metal oxides using new amphiphilic block copolymers

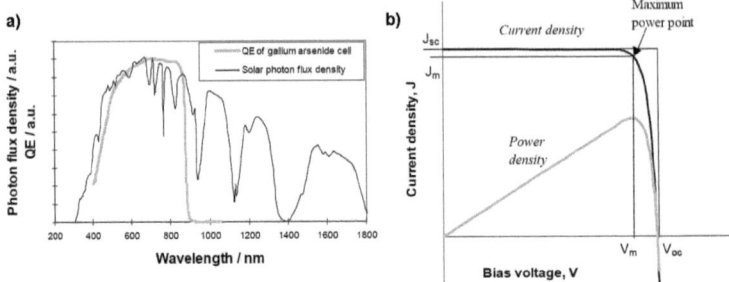

Figure 2.26: (a) Quantum efficiency (QE) of a GaAS cell compared to the solar spectrum. (b) Current voltage (black) and power-voltage (grey) characteristic of an ideal cell [136].

nanoparticulated film and the adsorption of the Ru-dye, the HTM was introduced into the mesopores. Finally, semitransparent thin gold layer constitutes the counterelectrode [132].

Motivation

In the last years many studies were carried out to improve the efficiency of the DSSCs. Most of these were devoted on preparing novel dyes and hole transfer materials, but less attention has been paid to TiO_2. In the present section the application of the formerly described titania mesoporous thin films as electronic conductor is faced. Besides the pure anatase phase, these systems result interesting because the employ of the sol-gel technique guarantees a more homogeneous crystalline network, enabling a better electronic percolation. Moreover, porous TiO_2 alike the PIB-PEO templated one, owning large pores, allows higher impregnation of the Ru-dye and a better HTM infiltration. Liquid and solid state DSSC were developed and tested within a collaboration with BASF AG at the Joint Innovation Lab - Organic Electronics in Ludwigshafen, Germany under the supervision of Dr. Felix Eickemeyer.

2.4.2 Characteristics of the photovoltaic cell

The main parameters which characterize a solar cell device are: the quantum efficiency (QE), the energy conversion efficiency (η_{CE}) and the fill factor (FF) [136]. The photocurrent generated by a solar cell under illumination is dependent on the incident light. The *quantum efficiency* (QE) is the probability that an incident photon will deliver one electron to the external circuit. QE depends on the absorption coefficient of the solar cell material, the efficiency of charge separation and the efficiency of the charge collection in the device. In Fig. 2.26a a typical QE spectrum in comparison with the spectrum of solar photons is shown. In operating regime the range of voltage in a solar cell goes from 0 to V_{OC}, where V_{OC} is the voltage at opened circuit. The cell *power density* is given by

$$P = JV \qquad (2.3)$$

where J is the density current. P reaches the maximum at the *maximum power point* (P_m). This occurs at the voltage V_m with a corresponding current density J_m (Fig. 2.26b). The *fill factor* (FF) represents the squareness of the power density curve, i.e. its ideal behavior, and is defined as the ratio

$$FF = \frac{J_m V_m}{J_{sc} V_{OC}} \qquad (2.4)$$

where J_{sc} is the short-circuit current density. The energy conversion efficiency (η_{CE}) is the power density at P_m, as a fraction of the incident light power density P_s

$$\eta_{CE} = \frac{J_m V_m}{P_s}. \qquad (2.5)$$

The energy conversion efficiency is also related to the FF,

$$\eta_{CE} = \frac{J_{sc} V_{OC} FF}{P_s}. \qquad (2.6)$$

2.4.3 Testing of dye sensitized solar cells

The efficiency of liquid and solid state DSSCs based on sol-gel TiO_2 semiconductors is compared with the nanoparticulate TiO_2 based ones. The following investigations show an overview on the potentialities of these new kind of devices, in order to trace the outline for future developments. The sol-gel systems present a thickness of 250 nm and were crystallized at a temperature of 550 °C.

2. Synthesis of mesoporous metal oxides using new amphiphilic block copolymers

Sample	V_{OC} (V)	J_{sc} (mA/cm^2)	FF	η_{CE} %	QE %
BASF	-	-	-	-	70
TP6_28	0.35	0.05	65	0.02	0.65
TP6_44	0.65	0.31	61	0.20	4.00
TP6_50	0.70	0.30	76	0.27	6.20
sputt-TiO$_2$ ECN	0.76	3.80	24	0.62	-
sputt-TiO$_2$ ECNb	0.70	2.62	24	0.44	-
solgel-TiO$_2$ ECN	0.78	0.97	43	0.33	-
sputt-TP6_44	0.72	1.32	28	0.27	-
solgel-TP6_44	0.68	1.10	42	0.32	-
solgel-TP6_44b	0.63	1.60	50	0.48	-

Table 2.4: Characteristics of liquid state (first row) and solid state (second row) dye sensitized solar cells. The samples sputt-TiO$_2$ ECNb and solgel-TP6_44b refer after 15 min of light exposition.

The preparation and the assembly of the solar cells is exaustively described in Appendix A.1.5.

The characteristics of the liquid-state DSSCs were studied as function of the template amount and are shown in table 2.4, first row. The labelling of the different samples follows the same criterion as for the TiO$_2$ films studied formerly. "BASF" is the liquid-state device developed by BASF AG used as reference. This sample owns highly crystalline, nanoparticulate, 10-μm-thick TiO$_2$ film as semiconductor. As one can see, the higher the template content is, the better is the cell efficiency. This result can most probably be due to an enhancement of the available surface area of the system, being the porous structure more compact and the accessibility of the pores higher. Thus, the grafting of the dye and the permeation of the electrolyte are facilitated. The efficiency of the sol-gel based solar cells, however, is still far away from the BASF ones. The main reason lays on the semiconductor's thickness, being almost 50 times thinner (10-11 μm vs. 250 nm).

Solid-state DSSC were also object of investigation, see Tab. 2.4 second row. The systems studied differ from the kind of TiO$_2$ compact intralayer: sputtered

2.4.3. Testing of dye sensitized solar cells

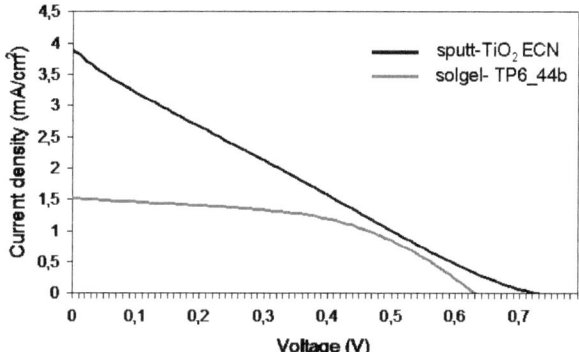

Figure 2.27: *JV* curve representation of the standard BASF (black line) and the sol-gel (grey line) solid-state DSSC.

(sputt-) or sol-gel coated (solgel-), and from kind of TiO_2 semiconductor: highly crystalline, nanoparticulate, 4-μm-thick (TiO_2 ECN) or sol-gel (TP6_44). As one can see, the choice of intralayer has a significant effect on the cell efficiency. In the case of the standard particulated systems higher efficiency is achieved with a sputtered intralayer, while sol-gel systems operate better with a sol-gel intralayer. A reasonable interpretation may be that the adhesion of the interfaces between layers owning similar (sputtered/nanoparticulated) or identical (sol-gel/sol-gel) properties is much more effective and guarantee a better electronic percolation. Interestingly, it can also be observed that after 15 minutes of light exposition, the performance of the standard BASF nanoparticulated cell decreases (sputt-TiO_2 ECNb), while for the sol-gel based increases (solgel-TP6_44b) and even overtakes it.

These alternative solid state sol-gel based DSSC showed remarkable results especially if one considers that the systems were not optimized for such applications. In fact, the minor thickness and the low crystalline fraction in comparison with the standard BASF cells, heavily affect the efficiency. Nevertheless, since sol-gel systems are characterized by the almost absence of grain boundaries, a better electronic conductivity through the semiconductor is permitted. The promising perspectives for such devices can be also highlighted by the kind of JV curve (Fig. 2.27) which is close to the ideal shape (cf. Fig. 2.26b). Furthermore, it is also noteworthy to point out, that the results of the sol-gel solid-state systems find correspondence with similar systems in literature, being in agreement with the results reported by Lancelle-Beltrame [137] and Wang [138].

2.5 Summary

In this chapter the templating behavior of new amphiphilic block copolymers with high hydrophobic contrast was presented. The chance to generate diverse systems has been revealed strategical to understand the templating features of these macromolecules. Especially in the preparation of nanocasted silica powders, it has been shown how mesoporous structures could be realized up to 70% wt. of PIB6000 owning very good structural order and large pore size. On the other hand the templating activity of PIB2300 resulted quite limited and mesoporous structures could be obtained up to 35% wt.. Interestingly, this behavior does not occur for the generation of silica thin film materials, where both polymers could generate ordered lyotropic crystalline phases up to 60% wt.. The main peculiarity is given by the PIB6000 systems which enlarge the pore size by increasing the polymer amount. The robustness of PIB-PEO templated materials was proved also by the generation of hierarchical materials by adding a molecular template, which lead to the formation of an organized alloy phase. The PIB-PEO mesostructure could maintain the well organized assembly even at considerable high content of added surfactant. The differences in the templating behavior smoothed out by the establishment of more complex systems then silica, such as titanium dioxide. In this case it was observed that by the enhancement of polymer amount more robust structures with thinner crystalline walls could be generated. The realization of well structured titania thin films found application in the establishment of TiO_2 based solar cells devices which showed very good and promising features for future developments.

Chapter 3

Microporosity determination of hierarchical mesoporous SiO$_2$ by in-situ SANS

3.1 Introduction

As could be seen in chapter 2 the preparation of mesoporous materials through the supramolecular templating of block copolymers generates ordered systems possessing narrow distribution of pore size and large surface areas. The latter is given by the geometrical interface between the micellar aggregtes and the metal oxide framework, which can be obtained through theoretical calculations, as shown in ref. [114]. Interestingly, it was found that the surface area values are much smaller than the experimental ones. This difference can essentially be ascribed to the presence of microporosity between the walls separating the mesopores. In many cases, the block copolymers used for nanocasting contain poly(ethylene oxide) (PEO) as hydrophilic block. Thus, as a consequence of the molecular templating of the PEO chains throughout the siliceous matrix, microporosity is originated [139–143]. Owning to the compatibility of the hydrophilic block with the ethanolic/siliceous phase, the PEO chains can be dissolved and act as molecular templates for the micropores. The templating behavior of the PEO chains can lead to a continuous network which connects the main mesoporosity, as shown in the works of Ryoo et al. on the preparation of platinum and carbon replicas of SBA-15 silica [144, 145]. In general, three possible scenarios for the embedding of

3. Microporosity determination of hierarchical mesoporous SiO_2 by in-situ SANS

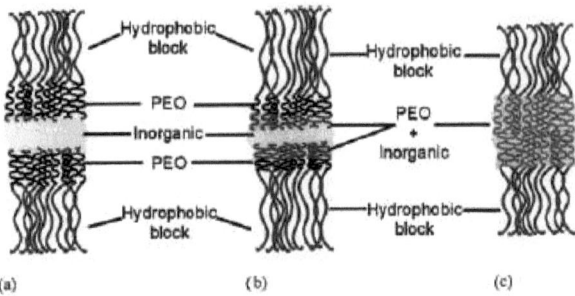

Figure 3.1: Different templating behavior of the PEO chains of block copolymers during a nanocasting process [114]

the hydrophilic chains in a silica matrix can be advised (Fig. 3.1) [114].

One possibility (Fig. 3.1a) is the formation of a pure PEO layer between the hydrophobic core of the micelle and the silica network. A partial dissolution of the PEO chains in the SiO_2 matrix can occur, thus leading to a templating action of the hydrated ethylene oxide chains and then to a certain degree of microporosity (Fig. 3.1b). Finally, the PEO chains are completely soluble in the inorganic phase and a pure templating process can be established (Fig. 3.1c). In addition, solid-state NMR studies corroborated this last finding revealing that the mobility of the hydrophilic chains is highly limited, since these groups are strongly anchored in the inorganic matrix [146]. It is straightforward that the removal of the molecularly embedded PEO chains through heat treatment will generate micropores of comparable size.

The main methods for the porosity determination in porous materials are represented by N_2 physisorption and SAXS analysis. Concerning the sorption measurements, the micropore volume is usually obtained by the comparison of plots (t-plot and alpha-plot) [23], which found application in several studies [142, 144, 147]. Neimark et al. set that these plots underestimate the total micropore volume [148]. Hence, consistent results are given by the application of a method based on the non-local density functional theory (NLDFT) (see section 1.1.2) [27, 149, 150]. The validity of this approach was confirmed by numerous studies [28, 29, 32]. Microporosity, expressed in terms of micropores' size, can be independently investigated by a well established method of evaluating SAXS curves, namely the

3.1. Introduction

chord length distribution (CLD) method (see section 1.2.3) [52, 59, 114, 151]. The application of this approach is particularly suitable in the case of low-ordered systems as the microporous one.

Although these techniques are widely applied, additional methods can help in understanding of sorption mechanisms. The combination of a physisorption experiment with small-angle neutron and x-ray scattering (in-situ SAXS/SANS-physisorption) represents an elegant way to describe the physisorption process in mesoporous materials (see section 1.3 for the theoretical principles). This approach revealed to details about porosity and connectivity, which cannot be obtained by the single techniques themselves [151–155]. Thus, since the micropores are responsible of the connectivity between the mesoporous cavities, this methodology is particularly suitable for their characterization [151].

In the present chapter the microporosity of two hierarchical porous silicas is analyzed and compared by means of in-situ SANS technique. The materials considered in this study are: PIB-IL 40 (in the following named as PIB-IL) and a similar material, KLE-IL, deeply characterized in section 2.3.2 and in ref. [109, 152], respectively. The main difference between these silicas lays on the kind of block copolymer adopted: PIB6000 for PIB-IL, and KLE [102] for KLE-IL. KLE, as already mentioned in section 2.2, is a diblock copolymer poly(ethylene-co-butylene)-b-poly(ethylene oxide). In addition to the mesoporosity, arising from the templating action of the spherical micelles of the block copolymers and from the cylindrical micelles of the ionic surfactant C_{16}mimCl (IL), these systems present a considerable amount of microporosity. Moreover, previous in-situ SANS studies on KLE-IL demonstrated that the majority of IL mesopores are effectively placed in the walls separating the KLE mesopores [109], thus enabling a hierarchical porous arrangement. This kind of material represents an ideal model system with respect to the study of connectivity in systems with pore hierarchy.

The investigation of microporosity in these complex materials using in-situ SANS methods can be motivated by the following aspects. First of all, it is important to quantitatively determine the fraction of micropores and their distribution, in order to evaluate the homogeneity of the materials. A semiquantitative approach, based on the analysis of the intensity changes in the SANS Bragg peaks [152], is used as alternative method to quantify micropore and small mesopore volume and to validate standard physisorption analysis. Furthermore, this method provides exhaustive information on the spatial location of the micropores and small IL mesopores. The sorption behavior and the size distribution of this kind of pores were also determined applying the CLD method to the analysis of the scattering curves. In-situ SANS measurements offer in this sense an interesting chance to relate diverse evaluation procedures and to verify standard sorption analyses. In addition, the comparison between two similar materials like KLE-IL and PIB-IL is noteworthy to understand the self-aggregation processes of different block copolymers by studying their microporosity as result of the templating action of the PEO chains.

3.2 Physisorption analysis

High precision nitrogen physisorption measurements were performed in order to compare and analyse the complex distribution of the porosity of these two hierarchical materials. In Fig. 3.2 the isotherms in linear and semilogarithmic coordinates (Fig. 3.2a and 3.2b) are shown. PIB-IL silica material exhibits a higher nitrogen uptake, with a larger hysteresis loop. The complete porosity determination was achieved by applying the NLDFT model. The pore size distribution (PSD) study (Fig.3.2c and 3.2d) evidences a similar trimodal distribution of pores in both materials. Besides the contributions of the PIB6000 and KLE pores at 17 nm and 13 nm respectively, two peaks centered at 2.5-3 nm and 1-1.5 nm are visible. These signals belong to the small IL mesopores and to the micropores, respectively. As one can see, in KLE-IL the micropores' contribution is much sharper and covers almost the double of the surface area than in the PIB-IL. The latter, on the contrary, is characterized by larger total pore volume and total surface area (Fig.3.2e and 3.2f). These and further sorption features are summarized in Tab. 3.1. It should be noticed that generally there is an evident, reasonable difference between the small volume of the micropores and the surface area which they occupy. On the contrary, the big mesopores present a relatively larger pore volume and smaller surface area contribution than the other types of pores. In addition, if one considers the BET surface areas it can be seen that these values are significantly larger than the ones obtained with the NLDFT method (Tab. 3.1). This finding is most probably due to overlap of sorption in the micropo-

3.2. Physisorption analysis

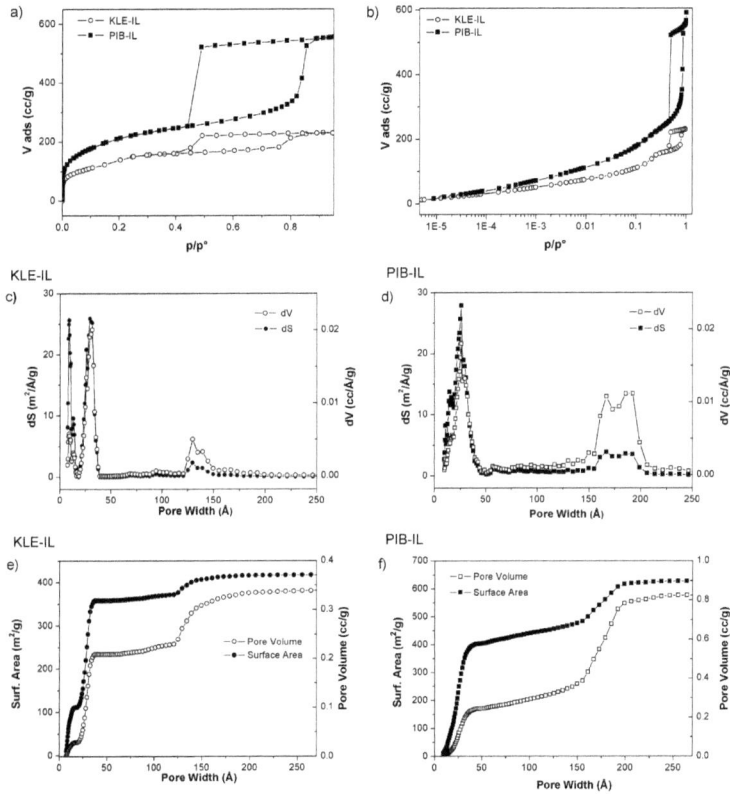

Figure 3.2: Nitrogen physisorption isotherms at 77 K of PIB-IL and KLE-IL in linear (a) and semilogarithmic plot (b). NLDFT differential pore volume and surface area of KLE-IL (c) and PIB-IL (d). NLDFT cumulative pore volume (white symbols) and cumulative surface area (black symbols) of KLE-IL (e) and PIB-IL (f). All the NLDFT analyses were calculated from the adsorption branch of the physisorption isotherms by applying the kernel of metastable adsorption isotherms based on a spherical/cylindrical pore model for the system nitrogen (77.4 K)/silica. It is important to underline that the lines connecting the data points serve just as "guides for the eye" and do not represent any fitting with physical meaning.

Property	KLE-IL	PIB-IL
mass (g)	0.116	0.055
surface area$_{BET}$ (m^2/g)	472	739
surface area$_{NLDFT}$ (m^2/g)	420	624
total pore volume (cm^3/g)	0.34	0.82
micropore volume (cm^3/g)	0.036	0.029
small IL mesopore volume (cm^3/g)	0.18	0.18
block copolymer mesopore volume (cm^3/g)	0.13	0.61

Table 3.1: Porous properties of KLE-IL and PIB-IL materials obtained through the NLDFT model. The values of the mass represent the material quantities used to perfom the standard physisorption analysis. The BET surface areas were calculated between $0 \leq p/p° \leq 0.1$.

res and in the IL mesopores in the range where the BET model was applied $0 \leq p/p° \leq 0.1$. All the more so, the BET method was not employed in the standard region $0.05 < p/p° < 0.3$ (see section 1.1.2), since for $p/p° > 0.1$ the filling of the IL pores occurs (see below), and the surface values cannot be completely trusted.

3.3 In-situ SANS data

Considering that this study is focused on the investigation of the micropores, the majority of the scattering curves were taken in the low pressure region (Fig. 3.3c,d). However, in order to comprehend the general pore filling behavior, several patterns were collected at the most representative points of the adsorption branch of the isotherms (Fig. 3.3a,b). The in-situ SANS experiment showed interesting and similar features concerning the porous structure and the general pore filling behavior of the KLE-IL and PIB-IL materials. At the void state ($p/p° = 0$; Fig. 3.4 A) the shape of both of the SANS curves for scattering vector $s < 0.25$ nm^{-1} was mainly determined by the block copolymer mesopores signals, namely by the Bragg peak reflection at $s \simeq 0.05$ nm^{-1} and by the form factor humps of the spherical mesopores between $0.1 < s < 0.2$ nm^{-1}. The broad maximum at $s = 0.3$

3.3. In-situ SANS data

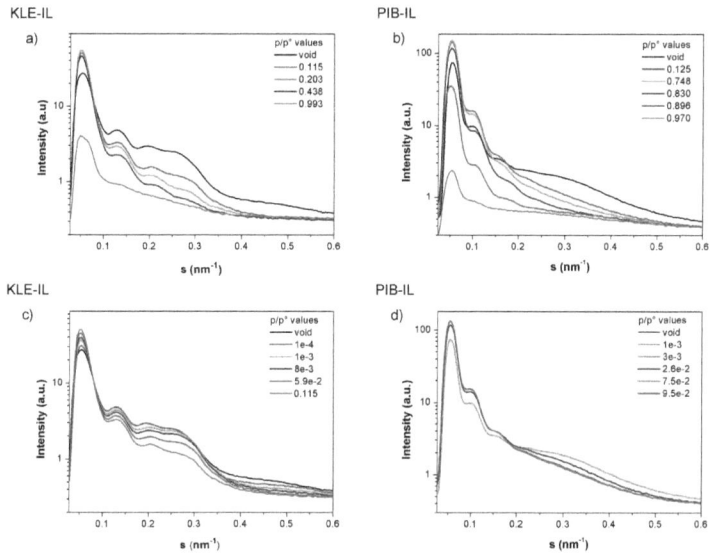

Figure 3.3: Scattering patterns of the materials collected at different pressures. Scans of KLE-IL and PIB-IL at representative points of the adsorption branch of the nitrogen isotherm (a, b) and in the region of the micropores filling (c, d).

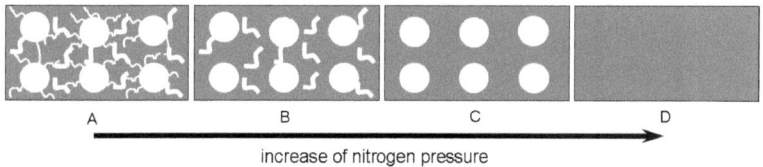

Figure 3.4: Structural construction of hierarchical silica illustrating the filling of the different pore types during increase in nitrogen partial pressure. Note that the idealized structural illustration in the first representation is to be proved in the present study.

nm^{-1} was assigned to the small IL mesopores, while the scattering at $s > 0.45$ nm^{-1} was supposed to originate from micropores. The basis for the following analysis of the porosity is the fact that the mutual arrangement of the micro-

3. Microporosity determination of hierarchical mesoporous SiO_2 by in-situ SANS

and mesopores in the silica matrix is reflected by distinct changes in the relative intensity of the different regions of the scattering curves when the pores start to get filled. From the void state (black curves) the patterns undergo constant increase of the Bragg peak intensity up to $p/p° = 0.2$ for KLE-IL, $p/p° = 0.125$ for PIB-IL, which corresponds to the filling of the micropores (Fig. 3.4 B) and of the IL mesopores (Fig. 3.4 C). Parallel to this, thanks to the contrast matching condition of N_2, the vanishing of the intensity of the corresponding pores' signals is also evidenced. Moreover, the enhancement of the Bragg peak is the result of an increase in average scattering length density of the silica matrix separating the spherical pores. These findings provide direct evidence of hierarchical organization of the pore network regarding the IL and block copolymer mesopores (see below). Finally, at $p/p° = 0.97$ the drastic decrease of the Bragg reflection reveals that also the spherical pores got filled (Fig. 3.4 D). The incomplete disappearance of the peak can be referred to the slightly imperfect matching of the scattering length density of N_2 and SiO_2, or to a very low fraction of inaccessible porosity. Since the SANS pattern are presented in logarithmic scale the residual intensity appears relatively large, but is at ca. 1.5-2 orders of magnitude smaller than in the evacuated state.

Furthermore, it is noteworthy to mention that the apparent absence of changes in the scattering curves of Fig. 3.3d up to $p/p° = 3 \times 10^{-3}$ is most probably due to a low resolution of the plot, which does not resolve the faint differences of the patterns.

3.4 Analysis of the in-situ SANS data by the CLD concept

The SANS curves were interpreted applying the CLD method, through an evaluation technique, which is exhaustively described in section 1.2.3. This mathematical approach provides a representation in real space of the scattering patterns, allowing to get direct information of the porous system evolution during nitrogen adsorption. In Fig. 3.5 the CLD plot of the nitrogen adsorption in micropores of both samples is shown. As one can see, up to pressures $p/p° < 10^{-2}$ the CLD presents a strong contribution at $r < 1$, which could be attributed to the average size of the micropores. The filling of the micropores, interpreted as the $g(0)$

3.4. Analysis of the in-situ SANS data by the CLD concept

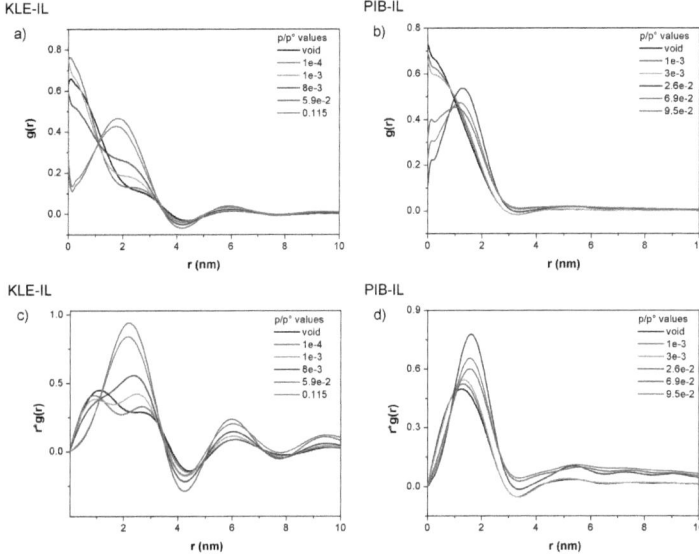

Figure 3.5: Chord length distribution plots $g(r)$ and their representation $r^*g(r)$ in the micropore filling region of KLE-IL (a, c) and PIB-IL (b, d).

decrease at $p/p° > 10^{-2}$, finds good agreement with the physisorption measurements (Fig. 3.2b). This large reduction of intensity in $g(0)$ implies that almost the totality of micropores were filled and then accessible to the nitrogen molecules. Similar to the SANS curves, the CLD has to be considered as a superposition of contributions from different kinds of pores. According to this statement, the CLDs of PIB-IL and KLE-IL possess different shapes, which depend on the material morphology. A comprehensive representation of the CLD is the plot of $r^*g(r)$ vs. r (Fig. 3.5 c,d) because the prevalent length-scales are obtained from the first momentum of $g(r)$ (see section 1.2.3). Under this assumption, in Fig. 3.5c the maximum at 1-1.2 nm of the void state curve can be referred to the KLE-IL average micropore size. The shoulder at ca. $r = 2$-3 nm belongs to an overlap of the IL pores and the wall separating them, while the minimum at $r = 4$ nm can be read as the sum of the average IL mesopore diameter and the corresponding pore wall. The confirmation of this last finding is the fact that the minimum

77

position remained unchanged with increasing filling, which is expected, since the average pore-to-pore distance of the IL mesopores has to stay constant. These results have been found consistent with the physisorption ones obtained by the NLDFT analysis (Fig. 3.2c). In the case of PIB-IL (Fig. 3.5d) the CLD is characterized by a main peak centered at 1.5-2 nm, which is obtained as a superposition of the contribution of the micropores size, IL mesopores size and IL pore walls. This effect can be explained with the help of the NLDFT pore size distribution analysis (Fig. 3.2 c,d). In the case of PIB-IL there is no clear separation between the micropores and IL mesopores. Furthermore, the corresponding pore sizes are respectively bigger (for micropores) and smaller (for IL mesopores) than in KLE-IL, thus promoting the merging of the single signals in the CLD plot. A validation of this assumption is given by the shift of the peak to higher r values when the micropores get filled, i.e. at $p/p° \geq 2.6 \times 10^{-2}$, meaning that the contributions of IL mesopore and IL pore walls get predominant. As for KLE-IL, the minimum at $r = 3\text{-}4$ nm can also be referred to the IL mesopore pore-to-pore distance, since it does not undergo changes during the sorption experiment. It deserves to be mentioned that the increase in the maximum around 2 nm for both CLDs for $p/p° > 10^{-2}$ has to be ascribed to the definition of CLD as a normalized distribution (see section 1.2.1): hence, the disappearance of the micropores resulted in a correspondingly higher contribution of the IL pores in the CLD.

In the case of KLE-IL materials the filling of the micropores and small mesopores was further analyzed by means of the Porod length l_p (section 1.2.1, Eqn. 1.13). The $r^*g(r)$ curves, the integrated area of which being l_p, were divided in two parts (Fig. 3.6a): a first one from 0 to 2 nm representing the micropore region, generating the mean micropore Porod length $l_{p,micro}$, and a second one from 2 nm, being mainly determinated by the superposition of the IL mesopores and the IL pore walls. The influence of the block copolymer pores can be neglected for this analysis. Fig. 3.6b shows that $l_{p,micro}$ possesses a value of ca. 0.7 nm which can be reasonably interpreted as the mean micropore size, considering an experimental error of ca. 10%. The imprecision in the estimation of the $l_{p,micro}$ is due to superposition of different sorption phenomena which take place at the same time. Nevertheless, the calculation of porous features, such as the micropore volume fraction (ϕ_{micro}), is still possible. This value can be obtained by means of $l_{p,micro} = 4\phi_{micro}(1 - \phi_{micro})V/S_{micro}$, namely Eqn. 1.21. The $l_{p,micro}$ value was

3.4. Analysis of the in-situ SANS data by the CLD concept

Sample	ϕ_{TOT}	$\phi_{IL+micro}$	ϕ_{IL}	ϕ_{micro}	S_{micro} %	$\otimes_{micro}$ [nm]
KLE-IL $_{nldft}$	0.42	0.26	0.22	0.036	27	0.8-1.2
PIB-IL $_{nldft}$	0.64	0.17	0.14	0.029	16	1.2-1.8
KLE-IL $_{cld}$	-	-	-	0.025	24	1-1.2
PIB-IL $_{cld}$	-	-	-	-	15	-
KLE-IL $_{bragg}$	0.41	0.25	0.18	0.069	-	-
PIB-IL $_{bragg}$	0.64	0.16	0.11	0.05	-	-

Table 3.2: Porosity features of the samples KLE-IL and PIB-IL obtained through NLDFT method (*nldft*), CLD approach (*cld*) and Bragg peak analysis (*bragg*). ϕ_{TOT}, $\phi_{IL+micro}$, ϕ_{IL} and ϕ_{micro} total volume fraction of the material, pore volume fraction of the small IL mesopores and micropores, pore volume fraction only of the small IL mesopores and pore volume fraction only of the micropores respectively, calculated through Eqn. 3.2. In the case of CLD method, ϕ_{micro} was obtained through Eqn. 1.21. S_{micro} % is the percentage of surface area occupied by the micropores. The symbol $\otimes$ means pore diameter.

obtained by the integration of the $r*g(r)$ plot (Fig. 3.6a), while S_{micro}/V could be calculated from the physisorption data. In principle the micropores surface area per volume is given by: $S_{micro}/V = \rho_{SiO_2} \phi_{TOT} S_{micro}/m$, where ρ_{SiO_2} is the mass density of silica ($\rho_{SiO_2} = 2.2$ cm^3/g), ϕ_{TOT} the total volume fraction obtained with the NLDFT method (Tab. 3.2) and S_{micro}/m the micropore surface area per gram (cf. Fig. 3.2e). Thus, the micropore volume fraction, calculated by $l_{p,micro}$ analysis ($\phi_{micro} = 0.025$), has been found in good agreement with the one calculated with the NLDFT kernel, see Tab 3.2.

The CLD can be also used to calculate the relative interface surface area of both samples as function of relative pressure. This value, expressed as percentage, was calculated from the changes in the Porod invariant Q (see Eqn. 1.11) and l_p. It should be noted that the achievement of Q was possible since the scattering patterns are expressed in terms of absolute intensity. Furthermore, because the volume of the sample hit by the neutron beam is unknown, i.e. the packing of the powder grains in the measuring cell is undefined, the Porod invariant allows the determination just of the relative and not absolute interface area (cf. Eqn. 1.19). Assuming that for KLE-IL and PIB-IL the majority of the micropores were

3. Microporosity determination of hierarchical mesoporous SiO_2 by in-situ SANS

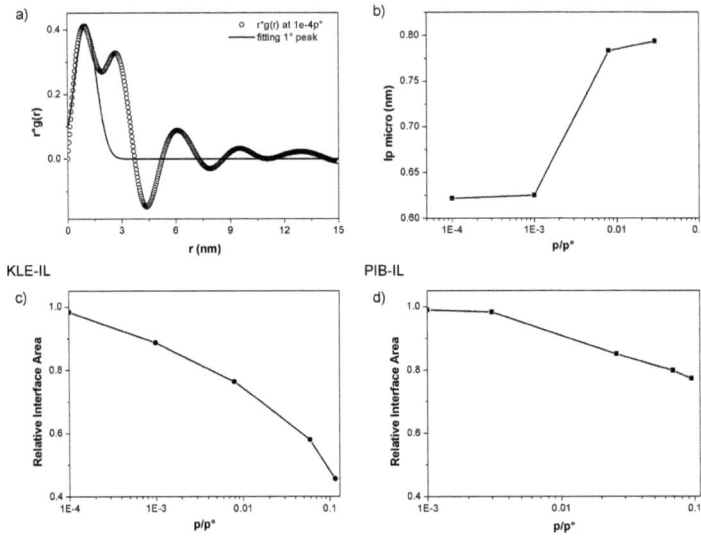

Figure 3.6: Separation of the micropores contribution in the CLD of KLE-IL at $p/p° = 0.0001$ (a); $l_{p,micro}$ plot in function of the partial pressure for KLE-IL silica (b); Relative interface area decrease in function of the partial pressure for KLE-IL (c) and PIB-IL (d).

filled at the corresponding relative pressures $p/p° = 0.008$ and $p/p° = 0.026$, a free interface area of 76% for KLE-IL and of 85% for PIB-IL can be estimated for these pressures (Fig. 3.6 c,d). This means that the area occupied by the micropores is of 24% and 15% respectively, values which are in good agreement with the NLDFT ones (Tab. 3.2). It can be excluded that the drop of surface area is significantly influenced by the sorption in IL mesopores, since no significant changes are shown in the CLD plots. In addition, it is interesting to observe how the Fig. 3.6c,d nicely qualitatively illustrates the continuous decrease in the surface area caused by layer formation at higher pressures.

3.5 Analysis of the Bragg peak intensity

The main goal of this evaluation method is to obtain accurate values for the volume fraction of micropores and IL mesopores direct from the change in intensity of the Bragg peak at ca. $s = 0.05$ nm^{-1} generated by the ordered block copolymer mesopores. Thus, this analysis allowed direct comparison with physisorption analysis by NLDFT. As it has been seen by the CLD, the micropores start to be filled at quite low pressures (ca. $p/p° = 10^{-3}$), subsequently at higher $p/p°$ also the small IL mesopores get filled. These effects can be observed by the increase in overall SANS intensity of the Bragg peak reflection already at small pressures (Fig. 3.3 c,d). In particular, the intensity of the Bragg maximum increases continuously during the filling of the micropores and IL mesopores. As pointed out in a recent work of Sel et al. [152], the changing in the relative, overall intensity of the SANS curves, was due to increase in the average scattering length density of the silica surrounding the block copolymer pores, owning to the filling of the micropores and IL mesopores. This change provides then information about the mutual arrangement of these smaller pores and their amount in the material. The volume fractions of the pores (micropores and IL mesopores) *only* relative to the volume of the walls, where the micropores and IL mesopores are located, separating the block copolymer cavities is given by

$$\tilde{\phi}_{micro} = \frac{V_{micro}}{V_{micro} + V_{IL} + V_{SiO_2}} \quad (3.1)$$

$\tilde{\phi}_{IL}$ and $\tilde{\phi}_{IL+micro}$, which are the volume fraction of the small IL mesopores and of the sum of the micropores and IL mesopores, respectively, are defined accordingly. These "partial" volume fractions can be obtained directly from the Bragg peak analysis (see below). On the contrary, the volume fraction of the pores referred to the *entire* volume, i.e including the volume of the block copolymer pores (V_{BC}), is given as

$$\phi_{micro} = \frac{V_{micro}}{V_{micro} + V_{IL} + V_{BC} + V_{SiO_2}} \quad (3.2)$$

$$\phi_{IL} = \frac{V_{IL}}{V_{micro} + V_{IL} + V_{BC} + V_{SiO_2}}$$

$$\phi_{IL+micro} = \frac{V_{IL+micro}}{V_{micro} + V_{IL} + V_{BC} + V_{SiO_2}}$$

$$\phi_{TOT} = \frac{V_{TOT}}{V_{micro} + V_{IL} + V_{BC} + V_{SiO_2}}$$

In order to calculate the ϕ values in the case of NLDFT analysis, the different pore volumes (V_{micro}, V_{IL}, V_{BC}) were obtained by Tab 3.1, while in the case of the Bragg peak analysis were calculated from Eqn. 3.1, opportunely modified in the case of ϕ_{IL} and $\phi_{IL+micro}$. It has to be noted that the value of V_{BC} for the Bragg peak analysis is the one achieved by NLDFT. V_{SiO_2} is given by the mass of analyzed material (cf. Tab. 3.1) and ρ_{SiO_2}. For the sake of clarity, in the Appendix A.5.2 a practical example of volume fraction calculation applying the Bragg peak analysis is given.

In order to separate the micropore from the IL mesopore contribution from the SANS pattern the pressures $p/p° = 0.008$ for KLE-IL and $p/p° = 0.026$ for PIB-IL were chosen. Even though in the low pressure range the micropores are completely filled, adsorption, i.e. formation of a monolayer, takes place in the pore walls of IL and block copolymer pores. However, in spite of this effect, the determination of microporosity is still possible. This formation of an adsorbed layer does not influence the analysis of the Bragg peak reflection in the block copolymer mesopores, because it practically does not affect the mean density of the surrounding silica matrix. More importantly, the maximum thickness at such low pressure is ca. 0.3 nm (see below), which is significantly smaller than the diameter of the spherical cavities. Such sub-monolayer formation has no appreciable effect on the form factor of the block copolymer pores (see Appendix A.5.1) and the respective absolute intensity of the Bragg reflection is hardly affected by this phenomenon. In the case of the narrow IL mesopores (diameter 3 nm) the formation of an adsorbed layer on their pore walls could in principle impact the Bragg analysis presented here. For instance, in the case of KLE-IL material, at $p/p° = 0.008$, where the majority of micropores have been filled, the statistical thickness of the adsorbed liquid-like N_2 layer is ca. 0.3 nm, based on a N_2 77 K reference isotherm obtained on amorphous silica LiChrosper Si-1000 silica [139], which is close to the monolayer thickness. This seems to be a reasonable value, since from the analysis of the KLE-IL silica one obtains a statistical thickness of 0.25 nm at a $p/p° = 0.008$. Hence assuming an adsorbed layer thickness of ca. 0.3 nm as upper limit, would result a situation that ca. 1/3 of the IL pore volume has already been filled. Despite this obstacle, it is still possible to extract the

3.5. Analysis of the Bragg peak intensity

micropore volume fraction from an analysis of the scattering data. As described in Appendix A.5.1, the micropore volume is given by the value obtained from the Bragg analysis for $p/p° = 0.008$ minus the pore volume occupied by the statistical layer in the IL mesopores. From the Bragg analysis of the KLE-IL silica it has been obtained $V_{micro+IL\,layer} = 0.07$ cm^3/g (see below), while from NLDFT V_{micro} = 0.03 cm^3/g. The difference (0.04 cm^3/g) is thus in perfect agreement with the volume expected for a statistical layer on the IL pores (see Appendix A.5.1). The same treatment is validated also for PIB-IL at the pressure $p/p° = 0.026$. For the sake of clarity, it should be noted that by accident some of the pore volumes have the same values of the volume fractions.

The "partial" micropore volume fraction ($\breve{\phi}_{micro}$) can be calculated through the equation:

$$(1 - \breve{\phi}_{micro}) = \sqrt{\frac{I_{void}(s)}{I_{micro}(s)}} \quad (3.3)$$

where $I_{void}(s)$ and $I_{micro}(s)$ are the maximum intensity of the Bragg reflection of the block copolymer pores in the void state and when the micropores are filled, respectively. Note that it is legitimate to use the absolute values of the Bragg peak maximum and not mandatory to use the integral intensities, because the shape of the Bragg reflection remains unchanged upon increase. In Appendix A.4.3 the achievement of Eqn. 3.3 is exhaustively described.

The micropore fraction between the block copolymer pores can be easily obtained then from Eqn. 3.3 knowing the intensity value of the Bragg peak maximum in the evacuated state and in the state where the micropores are filled. Similarly, applying this procedure to the maximum increase of the Bragg peak intensity up to the filling of the small IL mesopores ($p/p° = 0.25$), $\breve{\phi}_{IL+micro}$ can be easily calculated. The value of $\breve{\phi}_{IL}$ can thus be derived from the relationship $\breve{\phi}_{IL} = \breve{\phi}_{IL+micro} - \breve{\phi}_{micro}$. As mentioned above, from the single $\breve{\phi}$ it is possible to obtain the corresponding volumina of micropores and small IL mesopores (see Appendix A.5.2). Thus, the respective volume fractions referred to the entire volume of the material (ϕ) can be calculated from Eqn. 3.2. These latter values referred to micropores, IL mesopores and the total pore volume are shown in Tab. 3.2.

As one can see, the values of the volume fractions (ϕ), for both KLE-IL and PIB-IL samples, calculated through the Bragg peak analysis method, accord very well with the ones obtained through the NLDFT model. The only main difference is given in the calculation of the micropore volume fraction. In this case the values obtained with the scattering method are overestimated, with a consequent underestimation of the ϕ_{IL}. This as explained above, is due to the fact that when micropores get filled, the formation of an adsorbate monolayer in the small IL mesopores occurs. Thus, since these latter pores were assumed empty, their pore fraction results to be reduced. Even though this effect is present, the values can still be trusted, since they keep the same order of magnitude of the NLDFT ones.

Moreover, from these results important statements on the spatial location of the micropores an small IL mesopores can be derived. In fact, the Bragg peak analysis giving only information about the porous fraction of micropores and IL mesopores located in the block copolymer pore walls, and these ratios being in good agreement with the NLDTF values, which instead consider the sample in its own integrity, it can be concluded that the majority of both micropores and small IL mesopores are placed in the block copolymer pore walls, thus leading to highly homogeneous materials (see Fig. 3.4A).

3.6 General aspects of pore filling behavior

In section 3.4 it has been shown how it is possible to monitor the filling behavior of the micropres through the CLD concept. For the sake of clarity, in that section the attention was focused on the micropores and the small IL mesopores, ignoring the contribution of the block copolymer pores. Nevertheless, also the nitrogen physisorption mechanism in these latter cavities faces interesting features.

As one can see, the maximum at 12-13 nm in the enlargement of Fig. 3.7a represents the average size of the KLE block copolymer pores, according well with the NLDFT analysis. The variations of the maximum intensity can be explained considering that the CLD is a probability function. In this sense, at the void state no signals of the KLE pores can be detected, because the micropores and the small IL mesopores represent the majority of opened pores. Thus, filling these cavities at first, the scattering contrast between the matrix and the void KLE pores is enhanced ($0.06 < p/p° < 0.44$) and the intensity of the signal consequently increases. A similar sorption behavior for the spherical pores is shown also for the PIB-IL sample (Fig. 3.7b). In this case the large PIB mesopores contribution is

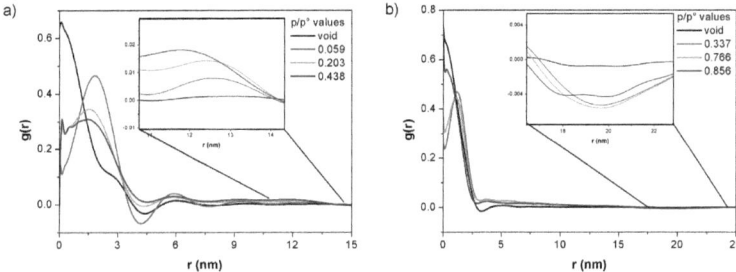

Figure 3.7: Chord length distribution plots with respective enlargement of the block copolymer pores signals of KLE-IL (a) and PIB-IL (b).

ascribed to the minimum at 20 nm which represents the pore-to-pore distance, i.e. the sum of the pore diameter and the pore wall. This result is in good agreement with the value obtained by the reciprocal of the Bragg peak maximum ($s = 0.053$ nm^{-1}), which gives $r = 19$ nm. It is noteworthy to mention that the small shift of the KLE pores signal in the range $0.06 < p/p° < 0.44$ is most probably due to the nitrogen multilayer formation. This effect can not be observed in the minimum of the PIB-IL silica, because the pore-to-pore distance is not affected by the nitrogen adsorption. Furthermore, an increasing of the $g(0)$ values by enhancing the pressure of adsorbed nitrogen can be observed. Considering that the CLD is nothing else than a mathematical representation of the scattering curve, and that all the micropores at $p/p° > 0.2$ are completely filled, these contributions have most probably to be ascribed to the scattering effect of liquid nitrogen.

3.7 Self-aggregation study of block copolymers by micropore analysis

The analysis of the micropores in KLE-IL and PIB-IL silica revealed to be also an appropriate way to understand the self-aggregation process of KLE and PIB6000 block copolymers. Smarsly et al. [106] already found that the length of the PEO chain of a block copolymer influences both the mesopore size and the amount of intrawall microporosity of the templated silica replica. In principle, if the hy-

3. Microporosity determination of hierarchical mesoporous SiO_2 by in-situ SANS

Property	PIB	PEO$_{PIB}$	KL	PEO$_{KL}$
ρ [cm^3/g]	0.84	1.13	0.88	1.13
Φ^{dft}	0.94	0.06	0.81	0.19
Φ^{th}	0.6	0.4	0.71	0.29

Table 3.3: Density values of the hydrophilic and hydrophobic blocks of PIB-PEO and KL-PEO block copolymers, at 25 °C. In the case of KL-PEO, being the hydrophobic chain a copolymer of poly(ethylene-co-isobutylene), PE$_{0.66}$-co-PIB$_{0.33}$, the entire density was obtained multiplying the density values of PE and PIB for the respective chain length fractions, χ. The volume fractions Φ^{dft} and Φ^{th} are obtained by means of Eqn. 3.4

drophilic block is significantly shorter than the hydrophobic one, a two phase system is formed, and the majority of PEO chains will be embedded in the silica matrix (Fig. 3.1c). On the other side, if the number of PEO units is comparable or larger than the hydrophobic ones, a fraction of the PEOs will be part of the micellar core (Fig. 3.1b), while the rest will form intrawall micropores. This representation of the block copolymer self-aggregation is also called "three-phase-model". In this sense, assuming that the microporosity of the analysed materials is originated only from the templating action of the PEO chains of the block copolymer, through the study of the micropore fraction, the micellization process of KLE and PIB6000 block copolymers can be studied. In order to clearly describe the behavior of the PEO chain in both block copolymers, in this section KLE and PIB6000 will be named as KL-PEO and PIB-PEO. Despite diverse hydrophobic blocks (cf. section 2.2) the main difference between these two amphiphiles is the length of the PEO chain, in fact KL$_{104}$-PEO$_{56}$ and PIB$_{107}$-PEO$_{100}$. Similarly, by using the chain length fractions χ, defined as the ratio between the length of one block and the total polymer length, we obtain: KL$_{0.65}$-PEO$_{0.35}$ and PIB$_{0.52}$-PEO$_{0.48}$. Thus, the comparison between the PEO micropore volume fraction obtained by means of the NLDFT model (Φ^{dft}_{PEO}) and the theoretical PEO volume fraction (Φ^{th}_{PEO}), which is supposed to generete only intrawall microporosity (Fig. 3.1c), gives information about the self-aggregation process.

These fractions are given by

$$\Phi_{PEO}^{dft} = \frac{V_{micro}}{V_{micro} + V_{BC}} \ ; \quad \Phi_{PEO}^{th} = \frac{\chi_{hyl}\,\rho_{hyl}^{-1}}{\chi_{hyl}\,\rho_{hyl}^{-1} + \chi_{hyb}\,\rho_{hyb}^{-1}} \qquad (3.4)$$

where V_{micro} and V_{BC} are the pore volume of the micropores and of the block copolymer pores obtained through the NLDFT model (Tab 3.1), respectively; ρ_{hyl} and ρ_{hyb} are the densities of the hydrophilic and hydrophobic block of the block copolymers respectively (Tab 3.3). χ_{hyl} and χ_{hyb} are the corresponding chain fractions of the hydrophilic and hydrophobic block.

As one can see, in the case of PIB-PEO, the PEO volume fraction which is embedded in the silica walls is much lower than the theoretical one. This means that most of the PEO chains of PIB-PEO polymer form a corona around the hydrophobic core of the micelle, contributing to the mesopore size. In fact, this finding would explain the larger cavities of PIB-PEO with respect to KLE-IL, even though both polymers possess hydrophobic blocks of the same length. On the contrary, in the KL-PEO polymer the experimental data are very close to the theoretical ones, meaning that the most of PEO chains behave as templating agents. The values of Φ^{dft} and Φ^{th} for PIB and KL blocks in Tab. 3.3 were obtained considering $\Phi_{hyb}^{th} = (1 - \Phi_{hyl}^{th})$.

3.8 Summary

In the present chapter it has been demonstrated how in-situ SANS experiments allow to study in detail the microporosity of complex hierarchical pores structures providing results which are in good agreement with the standard physisorption analyses. The application of the CLD approach revealed to be a unique chance to determine the pressures at which the filling of micropores and small mesopores occurs. Furthermore, the CLD being the representation in the real space of the scattering pattern, the morphology of the porous systems could be in depth investigated and especially its evolution during the filling process. The in-situ SANS approach revealed also to be a versatile technique, as it provides diverse modalities of data analysis. The study of the intensities variation of the Bragg maximum gave reliable results regarding the micropore and small mesopores fractions in the materials. Moreover, through this method it was possible to assess that the integrity of these pores were placed in the pore walls separating the block copolymer mesopores, thus proving the high homogeneity of the materials. The investigation of the microporosity in KLE-IL and PIB-IL was also remarkable in order to set the self-aggregation processes of PIB6000 and KLE block copolymers. With respect to KLE-IL, the microporous fraction in PIB-IL was found much smaller than the theoretical one, obtained considering a whole templating action of the PEO chains of the block copolymer. This finding has to be ascribed to a different micellization of the PIB-PEO block copolymer in which part of the PEO chains builds a corona around the hydrophobic core of the micelle.

Chapter 4

Vapor physisorption on hierarchical SiO$_2$ by in-situ SAXS/SANS

4.1 Introduction

In the present chapter, the adsorption at room temperature of organic adsorptives on hierarchical silicas will be investigated by means of in-situ SAXS/SANS experiments. The use of organic molecules is interesting both for the potential applications and for the structural characterization of these materials. The contrast matching with silica, fundamental precondition for these analyses, was realized with dibromomethane (CH_2Br_2) in the case of x-ray, and perfluoropentane (C_5F_{12}) in the case of neutrons (cf. Tab 1.1).

The importance of hierarchy for the industrial application of a porous material, especially with respect to separation process, was previously described (section 2.3.2). It is well known that the most of these processes take place at room temperature involving organic molecules. Hence, the utilization of the in-situ small-angle scattering technique, with respect to the standard characterization, is a unique chance for the direct study of the physisorption behavior by probing the porous texture of the material during vapors adsorption. Object of these analyses will be two different hierarchical silica, PIB-IL and KLE-IL, previously studied by in-situ scattering with respect to nitrogen adsorption (chapter 3). These materials are hierarchical in the sense that 3 nm worm-like pores (IL mesopores) are placed in the walls of larger spherical cavities (PIB, KLE mesopores). The PIB-IL material investigated in this chapter is taken from another batch compared to the one studied in chapter 3, but it presents the same structural and porosity features (cf. Appendix A.6.1). The employment of different experimental conditions than the standard ones, i.e. nitrogen at 77 K, helps in the understanding of the structural organization of the materials. In fact, the diffusion of relative large molecules like C_5F_{12}, being three times larger than CH_2Br_2 and N_2 (cf. Tab. 1.1) allows to study in detail the porous texture, in terms of connectivity and accessibility, especially considering the differences of these materials on the microporous scale, as assessed in chapter 3. For this purpose, a semi-quantitative analysis of the scattering curves by means of the chord length distribution method (cf. section 1.2.3) will help in providing valuable structural information.

Furthermore, these experiments are also relevant for a theoretical point of view, helping in the comprehension of unclear phenomena like the pore emptying behavior of ink-bottle pores. As it has been shown in section 1.1.2, such large spherical pores can be drained by so-called "pore-blocking" or "cavitation" mechanism, depending on the pore-neck size and on the experimental conditions of the measurement [33, 34]. Thus, a direct record of the pore emptying, by means of in-situ high precise SAXS/SANS experiments with different gases and at different temperatures, clearly identifies the kind of desorption process. In addition, the fitting of the desorption scattering curves by means of the Percus-Yevick (PY) model (cf. section 1.2.2), will help in finding a reliable relation between the morphology of the cavities and their draining pressure.

4.2 Vapor physisorption analyses

Vapor physisorption measurements show to be a precious tool for the study of the adsorption of organic molecules, even though highly precise porosity determination can not be carried out, due to the lack of appropriate sorption models for the corresponding adsorptives.

As one can see, the behavior of PIB-IL and KLE-IL silicas with respect to the perfluoropentane physisorption at 276 K is significantly different (Fig. 4.1a). The PIB-IL isotherm looks pretty similar to the nitrogen one in Fig. A.3a, while the

4.2. Vapor physisorption analyses

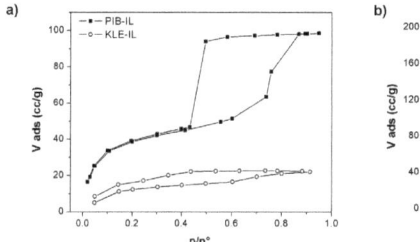
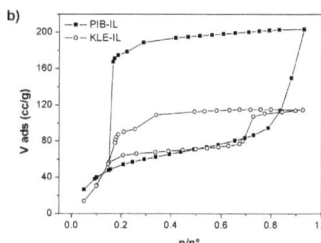

Figure 4.1: Vapor physisorption isotherms on PIB-IL and KLE-IL hierarchical silica of perfluoropentane (C_5F_{12}) at 276 K (a) and dibromomethane (CH_2Br_2) at 290 K (b).

Sample	$V_{p,C_5F_{12}}$ (cm^3/g)	V_{p,CH_2Br_2} (cm^3/g)	V_{p,N_2} (cm^3/g)
PIB-IL	0.77	≥ 0.64	0.82
KLE-IL	0.17	0.35	0.36

Table 4.1: Pore volume values calculated by means of the Gurvich law for the adsorption isotherms of C_5F_{12}, CH_2Br_2 (Fig. 4.1) and N_2 (Fig. A.3) at $p/p° = 0.9$. Note that PIB-IL is taken from another batch with respect to the one analyzed in chapter 3 but it presents the same features (cf. Fig. 3.2 vs. Fig. A.3).

KLE-IL one looks quite flattered, suggesting a scarce adsorption of the gas. In the case of dibromomethane (Fig. 4.1b), the most remarkable feature is the large hysteresis loop, shown by both of the isotherms, with the desorption occurring at very low relative pressures, $p/p° < 0.2$. This effect is intrinsic of thermodynamic properties of the adsorptive, such as mass density and saturation pressure.

Since the adsorption isotherms exhibit a plateau in proximity of $p/p° = 1.00$, is possible, by the amount of adsorbed volume, to calculate the pore volume of the material through the Gurvich law (Eqn. 1.2). For PIB-IL (Tab. 4.1) the pore volume is almost the same for N_2 and C_5F_{12}, in an experimental error range of 10%, suggesting that the porous texture is completely accessible to the gases flow. It should be noted that the value for dibromomethane is partially underestimated, since the isotherm, showing a capillary condensation step shifted to higher pressures, could not reach the plateau stage. In the case of KLE-IL the difference

4. Vapor physisorption on hierarchical SiO$_2$ by in-situ SAXS/SANS

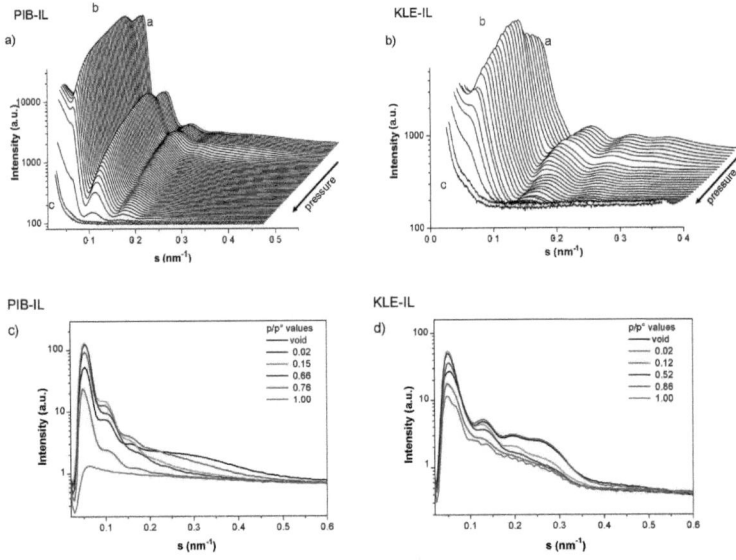

Figure 4.2: In-situ SAXS patterns of PIB-IL (a) and KLE-IL (b) during CH$_2$Br$_2$ adsorption at 290 K in constant flow conditions. In-situ SANS patterns of PIB-IL (c) and KLE-IL (d) during C$_5$F$_{12}$ adsorption at 276 K for significant relative pressures $p/p°$.

between $V_{p,C_5F_{12}}$ and the other two volumes, the former being less than the half is remarkable. This effect can most probably be due to the low accessiblity of C$_5$F$_{12}$ in KLE-IL matrix, and will be investigated in detail in the next section.

4.3 Adsorption in-situ SAXS/SANS data

A straightforward confirmation of the sorption results is provided by the adsorption in-situ scattering data.

In Fig. 4.2a,b the adsorption of CH$_2$Br$_2$ in PIB-IL and KLE-IL are shown. The increase in pressure is expressed by the direction of the arrow. Due to pore filling (cf. Fig. 3.4), the scattering curves undergo the same variation as in the case of N$_2$, as exhaustively explained in section 3.3. At the beginning, step $a - b$, the increasing of the intensity of the Bragg peak clearly indicates the micropores and

4.3. Adsorption in-situ SAXS/SANS data

small IL mesopores filling. Towards larger pressures, the multilayer formation and capillary condensation occur (step $b-c$). The latter phenomenon can be recognized by the vanishing of the Bragg peak. In c the total disappearing of the structural signals for both of materials shows the complete accessibility of the pore network to the CH_2Br_2 flow. This kind of experiment, performed at the synchrotron light, did not allow to precisely detect the reference pressure of each scattering curve, since the dibromomethane has been let flow continuously into the sample, without reaching equilibrium stages.

In Fig. 4.2c,d the neutron scattering curves relative to the C_5F_{12} adsorption in PIB-IL and KLE-IL are depicted. The scattering curves feature the same effects as in the case of CH_2Br_2 during the adsorption steps, i.e. micropore and IL mesopore filling, multilayer formation and capillary condensation. As also the ex-situ physisorption measurement suggested, the PIB-IL porous texture is entirely accessible to the gas flow. Owning to this, the determination of the pore volume of the IL mesopores and micropores ($V_{IL+micro}$) by means of the Bragg peak analysis is possible. Assuming that these pores are completely filled at $p/p° = 0.25$, the respective pore volume calculated by the method shown in section 3.5 (applying the Eqn. 3.3 and an approach similar to the one presented in Appendix A.5.2) are for CH_2Br_2: $V_{IL+micro} = 0.22$ cm^3/g and for C_5F_{12}: $V_{IL+micro} = 0.26$ cm^3/g. By comparison, the same values were obtained by the physisorption isotherms of CH_2Br_2 and C_5F_{12} (Fig. 4.1), at $p/p° = 0.25$ by using the Gurvich method (cf. Eqn 1.2). In the case of CH_2Br_2 was found $V_{IL+micro} = 0.19$ cm^3/g, while for C_5F_{12} $V_{IL+micro} = 0.24$ cm^3/g. The small varaince in the determination of the pore volume can be due to the difficulty to precisely identify the partial pressure at which the pores are filled and no multilayer formation in the bigger cavities occurs. Nevertheless, the values compared by the two methods are in good agreement, proving the reliability of the Bragg peak analysis in providing trustable porosity characterizations.

Differently from the case of PIB-IL, the scattering pattern of KLE-IL, at the saturation pressure ($p/p° = 1.00$), shows still a considerable fraction of void structure. The most probable explanation for the different behavior of the two materials towards the C_5F_{12} adsorption can be given by the analysis of the microporosity of the systems. As shown in the chapter 3, PIB-IL and KLE-IL possess different microporous structures due to the different block copolymers employed.

4. Vapor physisorption on hierarchical SiO$_2$ by in-situ SAXS/SANS

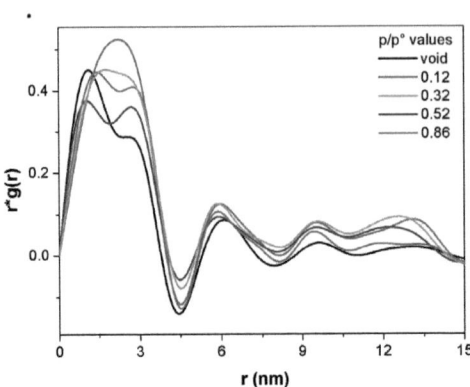

Figure 4.3: Chord length distribution functions for significant C$_5$F$_{12}$ relative pressure values $p/p°$ in KLE-IL silica.

KLE-IL has got a higher amount of micropores, but narrower in size than PIB-IL ones (cf. Tab. 3.2). In particular KLE-IL micropore dimensions (< 10 Å) are smaller than the C$_5$F$_{12}$ molecule, thus obstructing its diffusion. In theory, if one examines carefully the geometry of the gas molecule, being 10.8 Å long but only 4.4 Å wide (cf. Tab. 1.1), it could diffuse easily also in KLE-IL micropores. In fact, since these measurements are carried out at equilibrium conditions, the perfluoropentane molecule could arrange itself in order to minimize the energetic barriers and easily diffuse. However, this concept is valid only if it is assumed a pure cylindrical geometry of the micropores, which is not here the case. In fact, being the direct 1:1 replica of the solvated PEO chains of the block copolymer, they bear edges and corners, which represent a clear impediment to the gas flow. The confirmation of the micropore obstruction in KLE-IL silica for the C$_5$F$_{12}$ adsorption is given by the analysis of the CLD, in the form $r^*g(r)$, (Fig. 4.3). Looking at the curves after $p/p°= 0.3$, where the micropores and IL mesopores are supposed to be completely filled, it can still be recognized the distinctive values of the micropore and small IL mesopore size ($\approx$ 1 nm and 3 nm), meaning that part of these structures is still void.

The morphology of the micropores is not the only effect responsible for the

4.3. Adsorption in-situ SAXS/SANS data

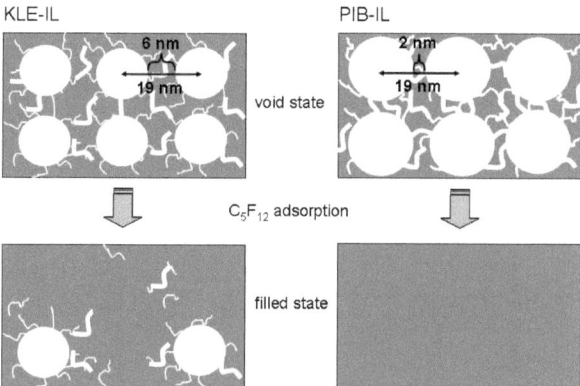

Figure 4.4: Schematic representation of the porous textures of KLE-IL and PIB-IL hierarchical silica at the void state and after complete filling with perfluoropentane

permeation of C_5F_{12}, since also the thickness of the pore walls plays an important role on the pore texture connectivity. Interestingly, both KLE-IL and PIB-IL are characterized by the same lattice parameter, i.e. pore-to-pore distance of the spherical mesopores, $a = 19$ nm. This value is obtained by the reciprocal of the Bragg peak which falls, for both materials, at $s = 0.053$ nm^{-1}. However, due to the significant difference of the spherical mesopore size, 13 nm and 17 nm for KLE-IL and PIB-IL respectively (Fig. 3.2c and Fig. A.3b), the materials are characterized by inverse pore wall thicknesses, i.e. 2 nm and 6 nm for PIB-IL and KLE-IL respectively, according to a (Fig. 4.4). In this sense, for PIB-IL possessing thinner walls, the connectivity between the lager mesopores is facilitated. Nevertheless, is important to point out that without the templating action of the IL mesopores, most of the spherical cavities would be inaccessible as demonstrated the porosity studies in section 2.3.2 (cf. Fig. 2.10d). Considering the size of the small mesopores (2.5 nm, Fig. 3.2c,d) and the pore wall thickness (2 nm), it is highly probable then, that the connectivity between the large cavities is established mostly by the IL mesopores, and the micropores just partially act as connectors. On the contrary, although KLE-IL silica presents a larger fraction of IL pores (cf. Tab. 3.2) the micropores still play an important role in the connectivity. In fact, since for the IL mesopores becomes difficult to connect directly two spherical cavities, due to the large pore walls (6 nm), the narrow micropores of KLE-IL act as bridges between IL and block copolymer mesopores. This statement is validated by the CLD plot in Fig. 4.3, where the remarkable signal of the small IL pores size at 3 nm for $p/p° > 0.3$ nm confirms the presence of void IL mesopores, which are not reached by the C_5F_{12}.

4.4 Desorption in-situ SANS data

The study of the in-situ scattering curves of the desorption branch of the isotherm represents an efficient way in the understanding of ambiguous phenomena like the pore emptying mechanism of ink-bottle pores. In particular, the main difficulty is to distinguish whether the pore draining process occurs via pore-blocking or cavitation. By standard sorption experiments this distinction becomes problematic, since the pressure of desorption p_d (Fig. 4.6b) is influenced not only by the geometry of the porous texture, but also by the kind of adsorptive and experimental temperature, as seen for the CH_2Br_2 isotherm (Fig. 4.1). In the present section, the emptying of PIB-IL hierarchical silica from several condensed adsorbates is investigated. In order to determine the influence of the experimental conditions, the studies on organic adsorptives at ambient temperature are compared with the one on nitrogen at 77 K.

In Fig. 4.5a-c the scattering curves relative to the desorption branch for C_5F_{12} at 276 K (Fig. 4.5a), CH_2Br_2 at 290 K (Fig. 4.5b) and N_2 at 77 K (Fig. 4.5c) are depicted. The pressure values, at which the curves are recorded, represent correctly the desorption steepness of the ex-situ isotherm in Fig. 4.1 and Fig. A.3a. This correspondence is not completely valid in the case of CH_2Br_2, for which the SAXS measurements were not carried out at equilibrium conditions. Nevertheless, the following considerations on its general desorption behavior hold true. However, the pressure values are shown just indicatively for helping the understanding of the emptying process. As one can see, as soon as the system overcomes the pressure of desorption p_d (Fig. 4.6b) the scattering intensity relative to the cage-like pores increases and the typical form factor pattern of a sphere can be recognized (red and green curves). This behavior means that the empty pores can be seen in the formalism of a diluted system of single spherical particles. Thus, it can be assumed that the pores empty randomly without any dependence from

4.4. Desorption in-situ SANS data

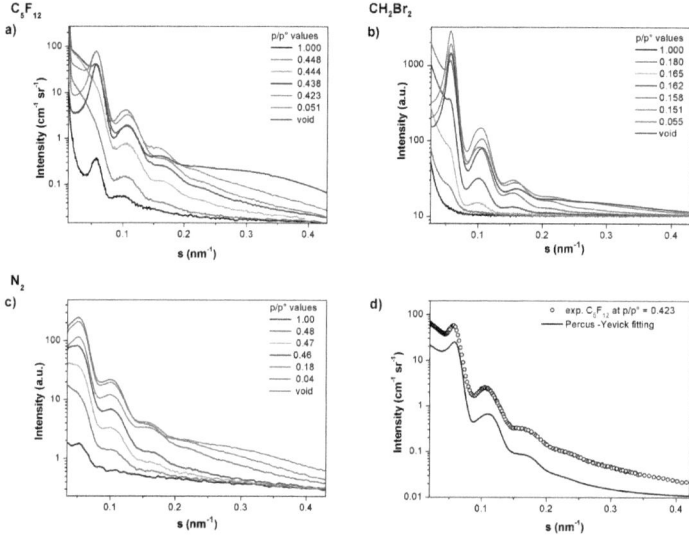

Figure 4.5: In-situ scattering patterns of the desorption process at selected pressures of C_5F_{12} (a), CH_2Br_2 (b) and N_2 (c) from PIB-IL hierarchical silica. Experimental desorption SANS curve of perfluoropentane at $p/p° = 0.423$ and its fitting by means of Percus-Yevick model (d). In (a) and (d) the intensity is expressed in [cm^{-1} sr^{-1}], since the C_5F_{12} desorption experiment was performed at the V4 line of the Helmholtz-Zentrum Berlin, allowing the acquisition of absolute intensities (cf. Appendix A.3.2).

their spatial position, i.e. independently form the lattice distribution (Fig. 4.6a A). This effect is further corroborated by the increase of $I(s)$ at small scattering vectors ($s < 0.05$ nm^{-1} Fig. 4.5a,b) as also observed in ref. [151]. This feature could be analysed just qualitatively since the fitting of the scattering curve (see below) was not possible within this region. The draining proceeding, the fraction of void spherical pores increases, so that the Bragg peak, which identifies the cubic structure of the block copolymer pores starts to become visible (Fig. 4.6a B). At the pressure of the closure point of hysteresis, p_{cp} (Fig. 4.6b), all the bigger pores are empty and the Bragg reflextion reaches its maximum (Fig. 4.6a C). As it can be observed, for all three gases, during the desorption hysteresis the scattering intensity relative to the small IL mesopores, i.e. the broad band

97

4. Vapor physisorption on hierarchical SiO$_2$ by in-situ SAXS/SANS

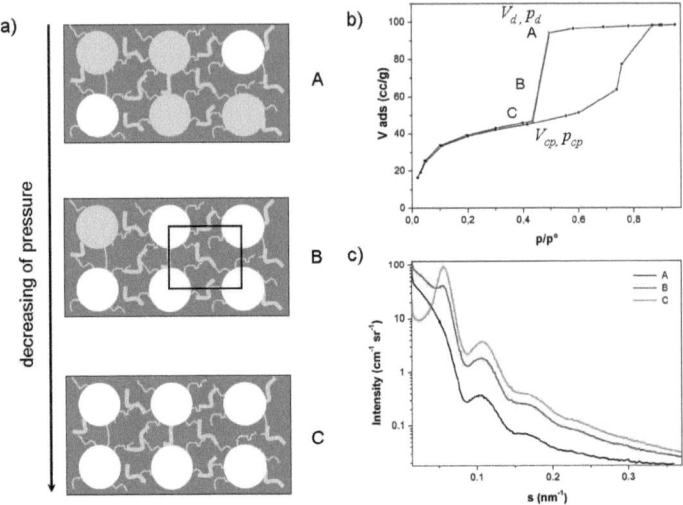

Figure 4.6: (a) Schematic illustration of the desorption mechanism in cage like pores of PIB-IL silica. A) Void pores with statistical distribution; B) Increment of the fraction of free pores and identification of the primary lattice organization (square); C) Complete draining of the cage-like cavities with the smaller pores being still filled. (b) Representation on a general physisorption isotherm of the investigated desorption region (red line) with the letters facing the main stages of desorption depicted on the left. The coordinates at the starting point of desorption are expressed by V_d, p_d, while at the closure point of the hysteresis by V_{cp}, p_{cp}. (c) Corresponding scattering patterns of the most meaningful desorption stages in PIB-IL silica.

at $s = 0.3$ nm^{-1}, did not increase. This means that these pores remained filled and the larger cavities drained by spontaneous evaporation of the adsorbate in the core of the pore. Hence, in PIB-IL desorption occurs through cavitation (see section 1.1.2). The smaller IL pores empty subsequently at very lower pressure $p/p° <0.1$. In addition, in the SANS pattern of C$_5$F$_{12}$ desorption at $p/p° = 1$, the small signals are most probably referred to inaccessible structures. Even if this residual intensity appears relatively large, one should consider that the SANS pattern are shown in logarithmic scale and the intensity is clearly smaller than in the evacuated state.

4.4. Desorption in-situ SANS data

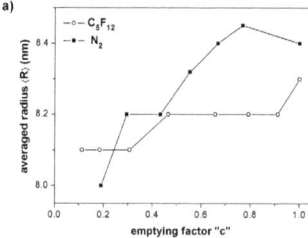
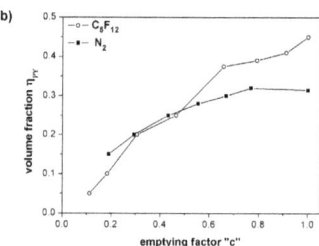

Figure 4.7: Plot of the averaged radius values, $\langle R \rangle$ (a), and of the volume fractions, η_{PY} (b), obtained by means of the Percus-Yevick model for C_5F_{12} and N_2 desorption in function of the emptying factor c.

Previously (cf. section 1.1.2), it has been reported that the cavitation mechanism in large cage-like pores (diameter bigger than $\approx$ 10 nm) occurs if the size of the pore neck is smaller than 4 nm, otherwise pore-blocking is the dominating process. This threshold was calculated in reference to the condition of nitrogen at 77.4 K [31], but is unknown for different adsorptives at different temperatures. These desorption studies have directly shown that the cavitation process occurs also for organic molecules like perfluoropentane and dibromomethane at ambient temperature conditions, in the case of porous networks bearing pore necks smaller than 3 nm.

The desorption in-situ scattering curves can also be fitted by means of the Percus-Yevick (PY) model in order to get precise information on the structure and morphology of the system during the emptying process. This approach is suitable in characterizing packings of spherical objects with local order like the mesoporous silica presented in this work, applying a model based on hard-sphere potential. The fitting of the experimental scattering curves provides the volume fraction of the spherical pores (η_{PY}) and their average radius ($\langle R \rangle$). As shown also by Smarsly et al. for similar system [151], the fact that the pores drain statistically without correlation between their size and position justifies the application of this analysis method. This approach could not be applied for dibromomethane, since the measurements were not taken at equilibrium conditions and the structural information given by the SAXS can not be correctly interpreted. The good

99

resolution in the PY fitting of the scattering curves is shown in Fig. 4.5d. The application of this model allows to monitor the increase of the volume fraction of the void spherical pores and the effect of the polydispersity during the draining process. These phenomena were analyzed respectively by plotting the structure parameter η_{PY} and the average radius $\langle R \rangle$, in function of the so-called emptying factor c (Fig. 4.7). The emptying factor represents the amount of void spherical pores and is given by

$$c = \frac{V_d}{V_d - V_{cp}} - \frac{V}{V_d - V_{cp}} \quad (4.1)$$

where V_d is the volume at the starting point of desorption and V_{cp} the volume at the closure point of the hysteresis (Fig. 4.6b). If one looks at Fig. 4.7a, the average pore radius increases, increasing the amount of voids both for C_5F_{12} and for N_2, meaning that the smaller pores empty first. This effect, due to the polidispersity of the sample, is consistent with the general statement of the modified Kelvin equation (Eqn. 1.3), for which: The lower the relative pressure is, the larger is the radius of the emptied pore ($p/p° \propto f(R^{-1})$). Accordingly to the draining process, the volume fraction of the empty mesopores increases (Fig. 4.7b). As one can observe, for C_5F_{12}, the final volume fraction of the free spherical pores is $\eta_{PY} = 0.45$ which is in whole agreement with the value obtained by the NLDFT method $\phi_{PIB} = 0.47$, obtained by $\phi_{PIB} = \phi_{TOT} - \phi_{IL+micro}$ (cf. Tab. 3.2). On the contrary, for N_2, even though the volume fraction increases, it does not reach the same value that in the case of C_5F_{12} (N_2: $\eta_{PY} = 0.31$). This lower result can be explained by the fact that the nitrogen measurements were carried out in a facility which did not allow high resolution analyses in terms of scattering intensity and small scattering vectors (cf. the sensitive areas of the detectors of the V1 and V4 line in Appendix A.3.2). For this reason a precise evaluation of η_{PY} is not possible in this case.

4.5 Summary

In the present section, in-situ small-angle scattering sorption experiments performed with organic vapors revealed to be suitable methods for the profound comprehension of the porous architecture organization of hierarchical silica. In particular, the different dimensions of the probing molecules established the opportunity to clearly determine the connectivity and the accessibility of the two similar material **KLE-IL** and PIB-IL, defining precisely the spatial location of micropores and small IL mesopores. From an applicative point of view, in order to facilitate the diffusion of large organic molecules throughout the material, a dense packing of large mesopoes has to be assured and the connection between them realized essentially by the the smaller IL mesopores. In this way the potential hindering of the micropores is avoided. Moreover, these in-situ SAXS/SANS-physisorption studies are also important for theoretical aspects. In fact, by the analysis of the curves collected during the desorption branch, it was possible to precisely identify the pore emptying mechanism of the spherical cavities. For different experimental conditions, it has been directly shown that the draining process occurs always by cavitation, i.e. it depends only by the thermodynamic conditions of the condensed adsorbate and not from the pore geometry. In this sense in-situ scattering experiments offer a smart alternative to the standard sorption measurements in the understanding of ambiguous phenomena like desorption from cage-like pores.

Chapter 5

Vapor physisorption on periodic mesoporous organosilica by in-situ SAXS/SANS

5.1 Introduction

Within this work the importance, for a porous silica material, to bear a hierarchical structure in order to increase its potential applications has been mentioned several times. Nevertheless, due to its chemical inertia, pure silica materials are limited in their range of functional properties. Therefore the recent discovery of porous organic-inorganic hybrid materials, and more precisely of periodic mesoporous organosilicas (PMO), has gained a lot of interest by the scientific community [156–159]. Instead of ordinary sol-gel precursors, special silsesquioxane compounds, characterized by an organic entity bridging between two alkoxysilane functions [(R'O)$_3$Si-R-Si(OR')$_3$], are used as building blocks for the porous network. The growing attention towards PMOs is given by the chance to incorporate organic moieties on the pore surface for potential purposes in catalysis, sensor technology and separation [80, 101, 160]. In particular, the object of investigation of this chapter will be materials possessing an enantioselective surface, capable to allocate a compound in a pure enantiomeric form, and thus particularly attractive for separation [161, 162].

Despite their manifold versatile structures, the porosity of PMO materials was not investigated as deeply as the pure silica counterparts. In the following,

5. Vapor physisorption on PMO by in-situ SAXS/SANS

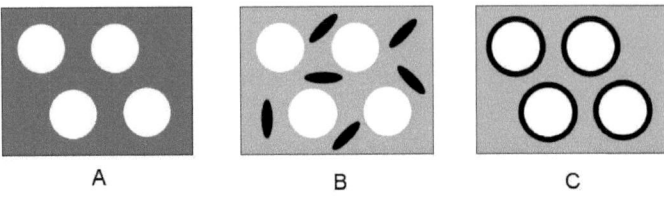

Figure 5.1: Illustration of the three possible scenarios for the potential distribution of the organic groups in PMOs. A) Homogeneous distribution throughout the matrix. B) Inhomogeneous distribution in the matrix, i.e. local enrichment in the matrix around the mesopores. C) Preferential enrichment at the pore surface.

the in-situ SAXS/SANS method is used for the characterization of the porosity and the structure of organosilica materials. For these purposes the use of organic probe adsorptives like dibromomethane and perfluoropentane, which fulfill the contrast matching conditions for silica with x-ray and neutrons respectively, were chosen. The advantages to perform a throughout structural characterization of the materials by means of in-situ scattering during adsorption of organic molecules at room temperature, is justified by several aspects.

Since these organosilica materials, bearing enantiomeric functionalities, are particularly suited for separation applications, it is of fundamental importance to study their sorption properties with respect to organic adsorptives at ambient temperature. The principle difference of organosilicas from SiO_2 materials is related to the presence of organic moieties, which grant a certain flexibility to the porous network that might induce a defined degree of microporosity. In particular, standard physisorption experiments revealed improper for porosity studies, because the low temperature conditions (77 K) might contract the flexible structure of the material, eventually closing the micropores. Furthermore, since the materials studied here underwent post-synthetic grafting with further organic functionalities (see below), some mesopores could be occluded and then not be detectable by physisorption. Through the in-situ scattering technique the structure of the materials is directly probed at each adsoprtion step, and the presence of inaccessible pores can be detected. Moreover, the chance to use organic adsorptives, like CH_2Br_2 and C_5F_{12}, with different molecular size (cf. Tab 1.1), leads to a better understanding of the pore connectivity towards different

5.1. Introduction

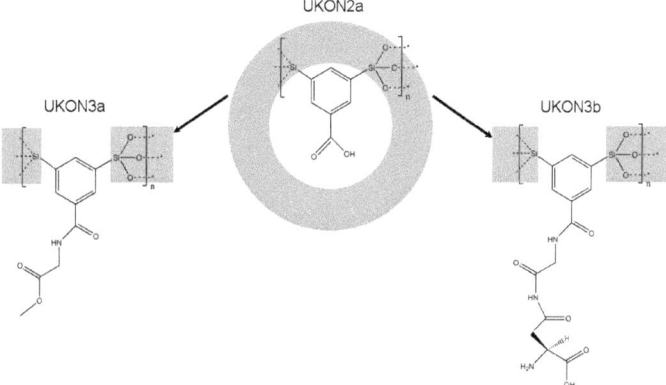

Figure 5.2: Schematical representation of the post-synthetic functionalization of the material UKON2a with alanine (Ala), resulting in UKON3a, or with the dipeptide alanine-asparagine (Ala-Asp), giving UKON3b.

fluids. The estimation of the pore size is also a matter of discussion. The analysis through nitrogen sorption possesses uncertainties because the interaction of nitrogen with the hybrid organic-inorganic surface is unknown. For this reason, theoretical models like BJH or NLDFT (cf. section 1.1.2) can not be considered accurate. In this sense, a quantitative analysis of the scattering curves can provide valuable structural information. By means of analytical methods like the chord length distribution and the Percus-Yevick approach, structural parameters like pore size, lattice parameter and pore volume, can be obtained for materials without high degree of mesostructural order.

In addition, these studies are relevant from a theoretical point of view. Structural investigation of PMO materials, especially regarding the displacement of the organic groups has not been carried out in depth so far. In fact, until now it was assumed, without any direct demonstration, that during the sol-gel reaction an ideal mixing of the silsesquioxane precursor molecules occurs, leading to homogeneous distribution of the organic moieties in the final material (Fig. 5.1A). However, different factors might be responsible for an inhomogeneous distribution. Since the organic groups and the hydrophilic siliceous groups are mutually incompatible to hydrolysis, domains (size ca. a few nanometers only) of several organic groups would form (Fig. 5.1B). In another scenario (Fig. 5.1C), it is possible that organic groups preferentially arrange around the micellar template, because of their dislike with silica. Hence, the investigation by the in-situ SAXS/SANS method, which employs fluids showing contrast matching conditions with the siliceous part of the hybrid materials, is of remarkable value in the comprehension of the PMOs organic moieties' location. By the analysis of the scattering curves at the filled state, in which no voids contribute to the scattering intensity, only the differences between the inorganic (siliceous) and organic part are detectable (in the case of perfect contrast matching conditions). In this way, if the materials are composed of a phase-separated system (Fig. 5.1B,C), the self-organized organic domains should be distinguishable in the scattering patterns. On the contrary, by a homogeneous distribution (Fig. 5.1A) the density differences are smeared and will not contribute significantly to scattering.

5.2 Materials investigated

As mentioned above, the materials object of the in-situ SAXS/SANS studies are organosilicas which present an enantioselective surface being able to separate compounds in a pure enantiomeric configuration. In fact, the chance to obtain stereochemically pure and specified molecules, e.g. amino acids, is considered a challenge of crucial importance. This target can be achieved for example by anchoring chiral building blocks onto the organic functionalities of the PMOs [162]. Hence, starting from a well established hybrid material, UKON2a, (Fig. 5.2) containing a bridging benzoic acid function along the pore walls [163], through the treatment with H_2N-AlaOCH$_3$ (Ala = alanine) and with H_2N-Ala-Asp(OCH$_3$)$_2$ (Asp = asparagine) the following materials UKON3a and UKON3b are obtained respectively. The detailed synthetic procedure is described in the ref. [163–166]

UKON3a and UKON3b were then characterized by means of SAXS, TEM and standard nitrogen physisorption methods (Fig. 5.3). For the sake of clarity, in Fig. 5.3a and Fig. 5.3b only the SAXS pattern and the TEM micrograph of UKON3b are depicted, being the UKON3a ones very similar. The scattering curve can be ascribed to a disordered mesoporous organization, with only local order. The Bragg maximum at $s = 0.15$ nm^{-1} identifies a lattice parameter, namely a pore-to-pore distance, $a \approx 6\text{-}7$ nm. The particular width of the signal identifies a

Sample	S_{BET} (m²/g)	V_{N_2} (cm³/g)	$V_{CH_2Br_2}$ (cm³/g)	$V_{C_5F_{12}}$ (cm³/g)
UKON3a	300	0.23	0.15	0.18
UKON3b	242	0.18	0.14	-

Table 5.1: Surface areas obtained by means of the BET model, and pore volume values for nitrogen, dibromomethane and perfluoropentane caclulated by the Gurvich law (Eqn. 1.2) from the adsorbed volume at $p/p° = 0.95$ for the sample UKON3a and UKON3b.

significant polydispersity of the cavities and distance distribution. From the TEM picture of UKON3b a disordered distribution of worm-like pores of ca. 4-5 nm can be deduced. A better understanding of the porosity of the two materials is achieved by the physisorption measurements. The N_2 isotherms (Fig. 5.3c) can be classified as H2 type (cf. section 1.1.1), characteristic of mesoporous systems with a relatively broad distribution of size and shape, validating the former results. The pore size analyses (Fig. 5.3d), obtained by means of the NLDFT model, show for both of materials a quite broad distribution of pores centered at 4 nm, with UKON3a possessing in addition a small fraction of bigger mesopores of 10 nm in size. According to the larger attached groups (two amino acids, i.e. alanine and asparagine), UKON3b has got lower surface area and pore volume, as shown in Tab. 5.1. It is noteworthy also to point out that by physisorption no micropores were detected.

5.3 Vapor physisorption analyses

Vapor physisorption measurements offer valuable information and a remarkable support to the in-situ scattering experiments. The isotherms of CH_2Br_2 adsorption at 290 K on the two PMO materials look pretty similar (Fig. 5.4a). As in the case of hierarchical materials (cf. section 4.2) the curves are characterized by a wide hysteresis loop shifted to lower relative pressure values. In fact, the hysteresis closes at $p/p° \approx 0.17$ instead at $p/p° \approx 0.4$, as in the nitrogen analysis. The sample UKON3a was additionaly investigated by C_5F_{12} physisorption at 276 K (Fig. 5.4b). As main feature, the isotherm shows a very narrow hysteresis loop at this temperature. The bare vapor physisorption data reveal already interesting

5. Vapor physisorption on PMO by in-situ SAXS/SANS

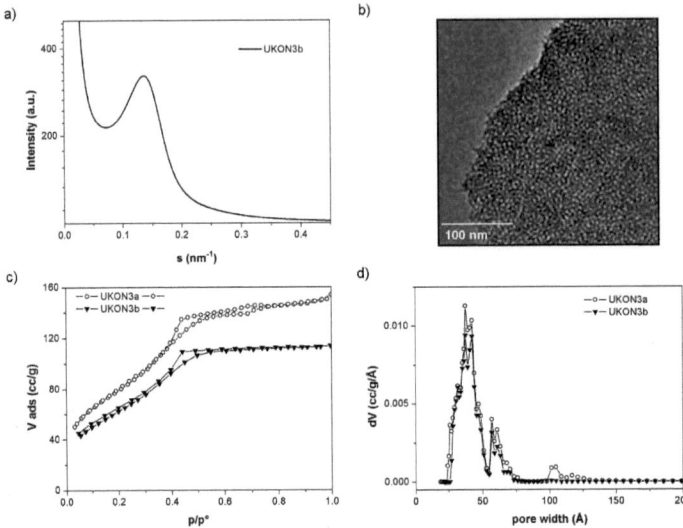

Figure 5.3: Small-angle x-ray scattering pattern (a) and TEM micrograph (b) of the sample UKON3b. Nitrogen physisorption isotherms at 77 K of UKON3a and UKON3b (c). Pore size distribution of the two hybrid materials from the adsoprtion branch calculated through the NLDFT method on the basis of a cylindrical pore model for the system nitrogen (77.4 K)/silica (d).

information about the material porosity. In particular, for UKON3a the porosity, expressed in terms of pore volume (Tab. 5.1), calculated for perfluoropentane is similar to the one for dibromomethane. Thus, even if perfluoropentane is three times larger than dibromomethane (cf. Tab 1.1), hindered adsorption can be excluded. The obstruction by this molecule is avoided, also because the mesopopores constitute here a through-connected network and are not linked through micropores as the hierarchical silicas previously described. Furthermore, since the physisorption behavior for vapor/PMO systems is not clearly understood yet, due to the unknown interactions of the organic adsorptives on hybrid organic-inorganic pore walls, is important to verify if the fluids are adsorbed following the BET or Langmuir theory. As one can see, the good fitting of the BET plot with the

5.3. Vapor physisorption analyses

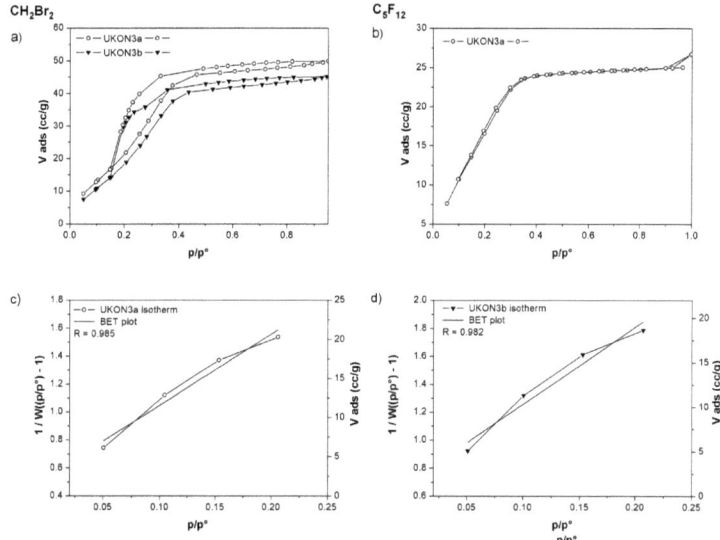

Figure 5.4: Dibromomethane physisorption isotherm of UKON3a and UKON3b at 290 K (a). Perfluoropentane physisorption isortherm of UKON3a at 276 K (b). Dibromomethane physisorption isotherms up to $p/p° = 0.22$ of UKON3a (c) and UKON3b (d) and respective BET plot. The term W in the y-axis is referred at the quantity of CH_2Br_2 adsorbed expressed in grams; R represents the correlation coefficient of the plot.

CH_2Br_2 isotherms of both materials (Fig. 5.4c,d $R > 0.98$, where R is the correlation coefficient of the fitting) ensures the presence of a BET adsorption behavior, implying the formation of multilayers of adsorbate on the material surface. By comparison, the Langmuir plots, showing worst correlation factor, are depicted in Appendix A.6.2.

5.4 In-situ SAXS/SANS data

In-situ small-angle x-ray scattering measurements on the sample UKON3a during dibromomethane adsorption are presented in Fig. 5.5a. The most significant feature in the in-situ SAXS curves is the decreasing in intensity of the scattering maximum (Bragg peak) upon increase in relative pressure of the adsorptive. By such in-situ scattering experiments, important features of the PMO porous architectures can be revealed. A first significant aspect is the constant shifting of the Bragg peak to smaller scattering values for the initial adsorption steps, i.e. until ca. $p/p° = 0.3$. This shift seems to correspond to an increase in the average pore-to-pore distance (real space) between unfilled mesopores, as a consequence of the fact that at a certain pressure the pores below a definite size are filled. Hence, since empty pores contribute to the scattering, the average distance between the remaining void pores increases, assuming that the size of individual mesopores does not correlate with their position. In addition, the intensity at smaller scattering vector ($s \approx 0.1$ nm^{-1}) rises up. This effect, already observed in section 4.4 and in ref. [151], is a consequence of the polydispersity of the mesopores. By further increase of the gas pressure ($p/p° > 0.3$), the scattering maximum becomes broader and its intensity decreases significantly. This finding is a consequence of the complete filling of the mesopores. A small hump at $p/p° = 1.00$ can still be recognized and is probably due to a small fraction of inaccessible pores or an imperfect contrast matching condition, as already experienced in the case of in-situ SANS experiment on nitrogen physisorption in section 3.3. The possible mismatch between the average electron density of CH_2Br_2 and the PMO, can be explained since the latter was "a priori" assumed as the silica one. However, the almost total decrease in scattering intensity at $p/p° = 1$, indicates almost perfect contrast matching, which is reasonable taking into account that the scattering intensity is dominated by the heaviest element in a sample (Si in this case).

Another important aspect which can be discussed by the analysis of the in-situ SAXS curves is the absence of micropores in the PMO matrix. The scattering intensity is very weak at $s > 0.35$ nm^{-1}. So, if micropores were present, one should observe an increase of the peak's intensity for the first adsorption steps ($0 \leq p/p° \leq 0.15$), due to the enhanced contrast between the void and the matrix, which is not the case for the present materials. The scattering observed at larger s is thus attributable to density fluctuations, creating a constant SAXS background.

5.4. In-situ SAXS/SANS data

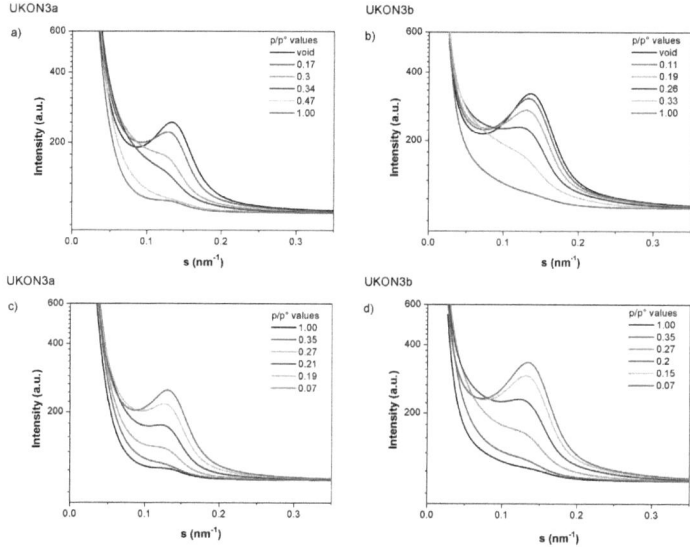

Figure 5.5: In-situ SAXS curves of dibromomethane adsorption in UKON3a (a) and UKON3b (b) at 290 K. In-situ SAXS curves of dibromomethane desorption in UKON3a (c) and UKON3b (d) at 290 K.

The physisoprtion behavior of the sample UKON3b (Fig. 5.5b) presents the same characteristics as the UKON3a, especially one can conclude that all the mesopores are accessible, as seen by the almost perfect contrast matching at $p/p° = 1.00$.

Furthermore, also the draining of CH_2Br_2 from the hybrid materials was investigated by in-situ scattering technique. The desorption patterns of UKON3a and UKON3b are depicted in Fig. 5.5c,d. As expected, a constant increase in the scattering intensity is observed, together with a shift of the main scattering peak maximum to larger scattering vector values. These experiments prove that the adsorption and desorption proceed reversible.

Physisorption in combination with in-situ SANS was also performed, using perfluoropentane on the sample UKON3a at 276 K (Fig. 5.6). The chance to use an adsorptive like C_5F_{12}, possessing a molecular size which is almost three times bigger than the CH_2Br_2 one, was expected to provide additional insights

111

5. Vapor physisorption on PMO by in-situ SAXS/SANS

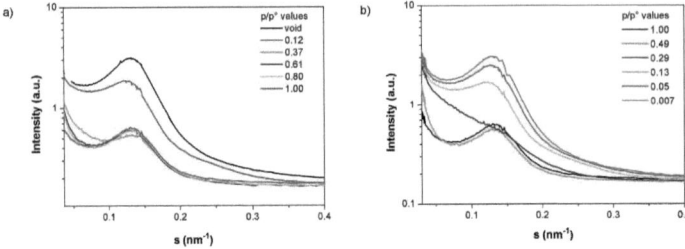

Figure 5.6: In-situ SANS patterns of UKON3a during prefluoropentane adsorption (a) and desorption (b) at 276 K.

into the PMO structure, since hindered sorption would occur, if the mesopores were connected through small micropores. As one can see, the SANS patterns look quite different to the SAXS data. Accordingly to the physisorption measurements, beyond $p/p° = 0.37$ no significant uptake was observed, expecting that all the mesopores were filled. However, the SANS curves reveal a pronounced scattering maximum even at further pressure than $p/p° = 0.37$. Since the physisorption experiment showed almost identical pore volume between N_2, CH_2Br_2 and C_5F_{12} (cf. Tab. 5.1), this remaining signal has to be attributed to mismatch of scattering length density (SLD) of perfluoropentane. In fact, assuming that the organic part of the material has the same mass density of the graphite ($\rho_C = 2.2$ cm^3/g), the corresponding calculated SLD is $\rho_{bC} = 7.33\times 10^{-6}$ Å^{-2} which is significantly different from the C_5F_{12} and silica one ($\rho_{bSiO_2} \approx \rho_{bC_5F_{12}} \approx 3.5\times 10^{-6}$ Å^{-2}, cf. Tab. 1.1). On the contrary, this difference is not so relevant in the case of x-rays, where the electron density of carbon and silica are not so different ($\rho_{eC} = 1.87 \times 10^{-5}$ Å^{-2}, $\rho_{eSiO_2} = 1.89 \times 10^{-5}$ Å^{-2}), thus explaining the almost perfect contrast matching.

5.5 Analysis of the in-situ SAXS curves

Before to focus the attention on the analysis of the SAXS curves, it is noteworthy to mention that the in-situ SAXS measurements were performed in constant flow of CH_2Br_2. Strictly speaking, as shown in chapter 4, being the system not in

5.5. Analysis of the in-situ SAXS curves

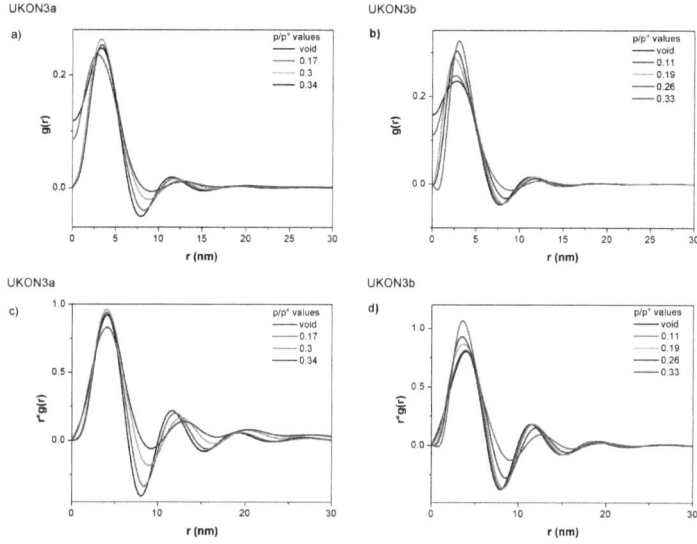

Figure 5.7: Chord length distribtion functions, $g(r)$, of dibromomethane adsorption on UKON3a (a) and UKON3b (b). Chord length distribution in the representation $r^*g(r)$ on dibromomethane adsorption of UKON3a (c) and UKON3b (d).

equilibrium, a reliable analysis of the SAXS curves in terms of equilibria could not be carried out. Nevertheless, one has to take in account that the structure of the present materials are less complicated, bearing a monomodal distribution of pores, and not a trimodal as in the case of the hierarchical silicas. The sorption process is therefore more straightforward, since the fluid does not have to diffuse in a complex porous texture possessing different pore necks. Hence, the in-situ SAXS curves, presenting only few scattering features can be quantitatively interpreted. Furthermore, the gases flow was kept constant for all the samples and pressures so that the data can be compared.

The in-situ SAXS curves of the dibromomethane physisorption on UKON3a and UKON3b were analysed by means of the chord length distribution approach. Interestingly, the value of $g(0)$ in the CLDs of the two samples (Fig. 5.7a,b) falls to zero for $p/p° > 0.2$. Generally, a positive value of $g(0)$ is caused by small

5. Vapor physisorption on PMO by in-situ SAXS/SANS

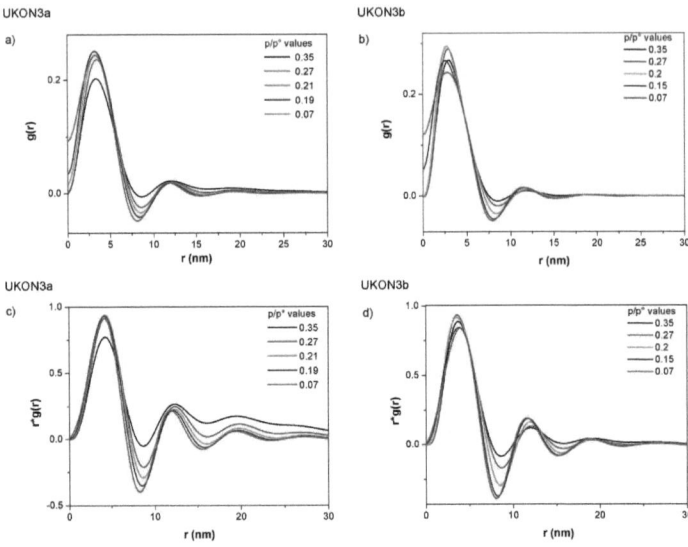

Figure 5.8: Chord length distribtion functions, $g(r)$, of dibromomethane desorption of UKON3a (a) and UKON3b (b). Chord length distribution in the representation $r^*g(r)$ of dibromomethane desorption on UKON3a (c) and UKON3b (d).

micropores and angularity of the pore system [59, 151]. The very small value of $g(0)$ indicates in this case the presence of a small degree of a certain angularity, most probably, from the irregular pore surface due to the grafting of the organic functionalities. The presence of micropores can indeed be excluded, since it is not observed neither in N_2 sorption analyses nor in the scattering patterns. The study of the morphological variations during the sorption experiment can be better appreciated by the analysis of the $r^*g(r)$ plot in Fig. 5.7c,d. The main peak at 4 nm represents a superposition of the contribution of the average pore size and the thickness of the pore wall. The following contributions at 8 nm and 12 nm correspond consequently to the pore-to-pore distance, and the chord penetrating two intraphases, i.e. two pore and one pore wall, respectively (cf. Fig. 1.6). A confirmation of these statements is given by the evolution of the $r^*g(r)$ curves during the adsorption process. It can be clearly seen that the pore-to-pore

5.5. Analysis of the in-situ SAXS curves

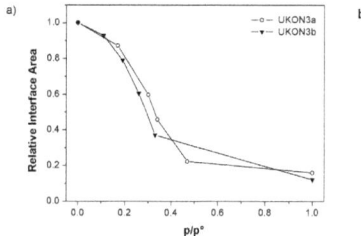
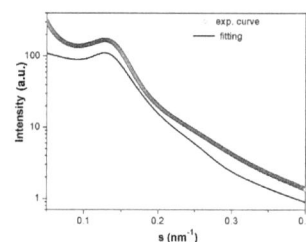

Figure 5.9: (a) Relative interface area plot as function of the relative pressure of CH_2Br_2 for UKON3a and UKON3b. (b) Experimental curve (circles) and PY fitting (bold line) of UKON3a at $p/p° = 0.17$.

distance of the system, i.e the minimum at 8 nm, increases by increasing the pressure. This has to be expected in fact, if holds true the assumption that the pores fill without correlation with their position. The peak at 4 nm, given by the sum of the contributions of pore size and pore wall, remains instead unchanged. This effect is given by the mutual annihilation of the changes of the pore wall and pore size signals during the physisorption process. In fact, since multilayer formation in the pores occurs, see section 5.3, the contribution of the pore walls shifts to larger values (pore wall and adsorbed dibromomethane layer), while, at the same time, the one of the size of the void pores reach lower values [151]. As a consequence, the superimposed maximum, sum of both contributions, does not change its position. Since the peak at 4 nm expresses also the pore size of the systems, the CLD analysis is in good agreement with the sorption experiments. Thus, even if the systems are constituted by a hybrid organic-inorganic texture, for the physisorption analyses the adsorbate/adsorbent interactions can be approximated to those of pure silica. The CLD plots of the desorption branch are the inverse of the adsorption (Fig. 5.8). Looking at the $g(r)$ representation, the $g(0)$ values increases by decreasing the pressure and reach almost the same value as in the void state. Accordingly, also the minima in the $r^*g(r)$ plot, referred to the pore-to-pore distance, decrease to smaller values. In the case of perfluoropentane, the calculation of the CLD is not worthy because of defective contrast matching.

The changes in the scattering intensity patterns allow to calculate the relative

UKON3a p/p°	η_{PY}	R_{PY} [nm]	σ_R	$\langle R \rangle$ [nm]
void	0.36	3.65	0.62	2
0.17	0.32	3.7	0.65	1.9
0.3	0.25	3.85	0.7	1.85
0.34	0.22	3.85	0.72	1.8
UKON3b p/p°				
void	0.34	3.6	0.6	2
0.11	0.34	3.7	0.6	1.9
0.19	0.32	3.7	0.62	1.8
0.26	0.28	3.75	0.65	1.78
0.33	0.22	3.8	0.68	1.75

Table 5.2: Analysis of SAXS data during dibromomethane adsorption by the Percus-Yevick model for hard-discs.

interface surface area as function of the relative pressure, by means of the calculation of the Porod Invariant Q (cf. Eq. 1.11). As one can see (Fig. 5.9a), for both samples the surface area mostly decreases until $p/p^\circ = 0.4$ according to the sorption analyses. Interestingly, a further evidence of the absence of microporosity in the systems is given by the smooth fall of the surface area for $0 \leq p/p^\circ \leq 0.1$. In fact, if micropores were present, a significant decrease in relative interface area should be noted at small relative pressure, as shown for hierarchical systems in Fig. 3.6c,d.

The scattering curves of the dibromomethane adsorption could also be analyzed by means of the Percus-Yevick model, since the materials present well defined mesopores, the size of which is probably not correlated with the position in the space (cf. behavior of the minimum at 8 nm in the CLD anaylsis). In this case, since the pores possess cylindrical, i.e. worm-like, form, the scattering patters can be described by using a model of a hard-discs [47, 151, 167]. The mathematical expressions of the form factor and lattice factor for this modified version of the Percus-Yevick model are presented in Appendix A.4.2. The good matching between the experimental data and the PY fitting is shown in Fig. 5.9b. As main advantage, this procedure provides meaningful physical parameters like the volume fraction η_{PY}, the PY radius R_{PY}, which represents half of the pore-to-pore distance, the average pore radius $\langle R \rangle$ and its variance σ_R which represents the polydispersity of the pore size (Tab. 5.2). By the analysis of the SAXS adsorption curves it can be seen that during the sorption process the pore-to-pore distance, R_{PY}, increases, while the average pore size, $\langle R \rangle$, is slightly decreased due to the formation of multilayer of adsorbate in the pores. This trend and the respective calculated values are in very good agreement with the results of the CLD, providing the reliability of both methods in the interpretation of the SAXS data. Furthermore, the polydispersity, σ_R, of the systems is enhanced at higher pressure, thus precisely explaining the rising of the intensity at small scattering vectors in Fig. 5.5a,b. In addition, the volume fraction of the mesopores η_{PY} decreases. In particular, the η_{PY} values at the void state are in quite good agreement with the volume fraction obtained by the sorption analysis ($\phi_{UKON3a} = 0.33$ and $\phi_{UKON3b} = 0.28$). These latter results were obtained by Eqn. 3.2, adopting the pore volume calculated for N_2 in Tab. 5.1.

5.6 Summary

In the present chapter the pore structure of two PMO materials was deeply studied by means of in-situ SAXS/SANS physisorption using organic vapors as adsorptives. Through these experiments, contrast matching was observed using CH_2Br_2 and SAXS for both samples, evidencing the complete accessibility of the materials. Furthrmore, this finding proves that the pore walls constitute a homogeneous matrix in terms of electron density, and thus of mass density. It can be therefore concluded that the organic moieties are randomly distributed in the material according to the representation of Fig. 5.1A. In-situ experiments also established the absence of microporosity, being the pores exclusively originated by the templating action of the surfactant and completely accessible to larger molecule like C_5F_{12}, as the physisorption measurements assessed. The quantitative analyses of the SAXS curves highlighted the good agreement with the nitrogen sorption experiments regarding the porosity determination. In this sense, it can be assumed that the interactions of the adsorbate with the material surface are of the same kind that the one of silica, validating the use of mathematical models like NLDFT for the material porosity characterization.

Conclusion

The present work dealt with the preparation of mesoporous metal oxide systems by means of novel block copolymer templates and the development of original structural characterization methods.

In the first part of the work was shown that PIB-PEO block copolymers, i.e. PIB6000 and PIB2300, bearing different hydrophilic blocks, generate robust and highly porous metal oxide textures characterized by large pores (15-25 nm) of spherical form. The synthesis of silica powder materials pointed out the templating differences between the two amphiphiles. In fact, with respect to PIB2300, PIB6000, characterized by a higher hydrophobic contrast, could form lyotropic liquid crystal phases up to 70% wt. in solution. Regarding the preparation of mesoporous thin films of SiO_2 and TiO_2, both polymers showed almost identical templating behavior. In the case of silica, ordered mesoporous systems could be established between 12% and 60% wt. of polymer. The spherical pores of ca. 20 nm in size organize following a distorted face centered cubic (FCC) packing, and for PIB6000 cavities up to 28 nm in size by high template contents could also be formed. Titanium oxide thin films were studied as function of amphiphile amount and crystallization temperature. All systems showed a distorted body centered cubic (BCC) arrangement of the pores, the organization and robustness of which increased with the adding of template. These effects were achieved by the formation of smaller crystallites and pores due to denser packing of the template micelles. The pore size dropped in fact from 20 nm to 16 nm, while the size of crystallites from 14 nm to 7 nm for the systems with 28% wt. and 50% wt. of polymer, respectively. These sol-gel mesoporous TiO_2 films were also tested as semiconductors in dye sensitized solar cell (DSSC) devices showing promising results in comparison with the nanoparticulate systems. The absence of grain boundaries allows better conducting conditions avoiding shortcuts, even though the thickness of the films was too small to ensure comparable efficiencies with the commercial devices. Furthermore, the remarkable templating behavior of the PIB-PEO block copolymers was shown in the preparation of hierarchical silica powders. The high hydrophobic character of these templates allows to obtain very homogeneous materials with trimodal distribution of pore size and large porosity.

Due to the structural complexity of these latter systems their characterization could not be exhaustively carried out by means of standard analysis approaches. The employment of the in-situ SAXS/SANS-physisorption technique provided thus an exclusive possibility to study in detail the porosity and its organization in these hierarchical materials. The remarkable reliability of this technique is confirmed for instance in providing alternative ways for the characterization of microporosity with respect to the bare nitrogen physisorption methods. The analysis of the scattering curves, collected during the nitrogen physisorption process, by means of the chord length distribution method and a semiquantitative analysis of the Bragg peak intensities of the scattering curves, allowed to precisely quantify the micropores' volume fraction, size and spatial distribution in PIB-IL and KLE-IL hierarchical silicas.

The manifold application perspectives of this analysis technique are given by the chance to work with different adsorbable gases and temperature conditions. In this work it has been shown in fact that in-situ SAXS/SANS experiments of organic vapors physisorption at room temperature conditions could elucidate fine structural differences in PIB-IL and KLE-IL. The possibility to adsorb organic molecules of different size, i.e. C_5F_{12} and CH_2Br_2, clarified the role of IL mesopores and micropores in the connectivity of these materials. Furthermore, in-situ SAXS/SANS-physisorption technique was found to be significantly suited for the explanation of gas desorption behavior in mesoporous materials. In fact, for the first time it was possible to recognize and directly study the cavitation pore emptying mechanisms occuring in cage-like pores.

Interesting applications of this novel characterization technique are given also by the analysis of further systems like periodic mesoporous organosilicas (PMOs). In the last part of this work the porosity and structure of PMO materials, bearing grafted enantiomeric functionalities, were deeply invesitgated. The use of the in-situ SAXS/SANS-physisorption technique proved the complete accessibility of the porous network, which was not affected by the post-synthetic functionalization. Moreover, structural studies evidenced the homogeneous distribution of the organic moieties throughout the hybrid materials.

Appendix A

A.1 Synthesis strategies

A.1.1 Nanocasted silica powders

A set of different PIB-PEO template concentrations was prepared as follows (summarized in Tab A.1): 200 mg of organic silica precursor tetramethoxysilane (TMOS) was used for each sample, to which appropriate amounts of a 5% wt. ethanol/THF (3:1 vol.) solution of the block copolymer were added drop by drop. After the addition of 100 mg aqueous hydrocloric acid (pH = 2), the samples were treated with ultrasounds for 5 min, and then the solvent was evaporated under gentle vacuum in a rotating evaporator. The resulting gels were aged at 60 °C in a drying oven for 10 h. The silica casts were finally calcined in air at 550 °C for 5 h.

A.1.2 Nanocasted hierarchical silica powders

The hierarchical silica powder materials were prepared in the same way of the monomodal powders. The only difference lies in the introduction of different amounts (10% - 100% wt. with respect to the block copolymer template) of the ionic liquid $C_{16}mimCl$ directly in the 5% wt. PIB-PEO solution.

A.1.3 Mesoporous silica films

In a typical synthesis of mesoporous silica films 4 ml of ethanol/THF (3:1 vol.) solution containing the desired amount of polymer (see Table A.2 first row) was added to a 2 ml of A2* solution (see below for the A2* preparation). To this solution, 0.32 g of HCl 1M were added drop by drop. The resulting sol was stirred

A. Appendix

Sample	PIB-PEOs sol. [mg]	HCl [mg]	TMOS [mg]	Template [wt.-%]
P6_20/P23_20	395	100	200	20
P6_25/P23_25	526	100	200	25
P6_30/P23_30	677	100	200	30
P6_35/P23_35	850	100	200	35
P6_50	1540	100	200	50
P6_70	3680	100	200	70

Table A.1: Experimental composition of the set of silica powders prepared from PIB-PEO polymers (PIB6000 and PIB2300). The ratio of template in the material was calculated as mass of polymer/(mass of polymer + mass of SiO_2).

for 15 min before the use. Silica films were produced via dip-coating on silicon wafers substrates. Optimal conditions were given for 20-25% of relative humidity and a constant withdrawal speed of 6.5 mm/s. The as-prepared films were aged in oven at 80 °C and subsequently calcined at 550 °C for 4 h (heating ramp 5 °C/min) in a muffle oven.

Concerning the hierarchical mesoporous silica films the synthetic route follows the one of the monomodal systems except for the introduction of the ionic liquid template (20% wt. with respect to the block copolymer), which was added to the PIB-PEO solution.

A2* Solution. The prehydrolized silica precursor solution was obtained by mixing 61 ml of TEOS in 61 ml of ethanol, subsequently 4.9 ml of H_2O and 2 ml of HCl 0.07 M were added. The whole solution was stirred for 90 min at 60 °C and then stored in fridge.

A.1.4 Mesoporous titania films

The mesoporous titania thin films were synthesized as follows. To 600 mg of titanium tetrachloride 3 ml of ethanol and 500 mg of bidistilled water were added drop by drop. Subsequently a 3 ml ethanol/THF (3:1 vol.) solution containing the desired amount of polymer template (see Tab. A.2 second row) was added. The obtained sol was stirred for 3 h before the use. TiO_2 films were produced via dip-coating on silicon wafer substrates. Optimal conditions were given for 15-20% of

Sample	PIB-PEO [mg]	Template [wt.-%]
SP6_12/SP23_12	38	12
SP6_18/SP23_18	61	18
SP6_28/SP23_28	108	28
SP6_50/SP23_50	277	50
SP6_60/SP23_60	416	60
TP6_28/TP23_28	100	28
TP6_40/TP23_40	168	40
TP6_44[a]	200	44
TP6_50/TP23_50	250	50

Table A.2: Block copolymer amounts used for the synthesis of the mesoporous MO_2 films (M = Si, Ti). The ratio of template in the material was calculated as mass of polymer/(mass of polymer + mass of MO_2). [a]This sample was synthesized only for the use in solar cell devices.

relative humidity and a constant withdrawal speed of 6.5 mm/s. The as-prepared films were aged in oven at 80 °C and subsequently stabilized at 300 °C for 12 h (heating ramp 3 °C/min). The throughout calcination and crystallization of the matrix were achieved by heating the samples from room temperature to 550 °C or 600 °C (temperature ramp 5 °C/min). The samples were immediately removed from the oven when the desired temperature was reached in order to avoid the collapse of the mesoporous structure.

A.1.5 Preparation of the DSSC devices

Liquid-state DSSC devices

The liquid-state solar cells were prepared by dip-coating of conducting substrates of FTO glass following the procedure described in A.1.4. The mesoporous TiO_2 films were soaked with an ethanolic solution 0.5 mM of the Ru dye (Fig. A.1a) for 18 h. The samples were washed with ethanol and let dry for 2 h at 40 °C.

A. Appendix

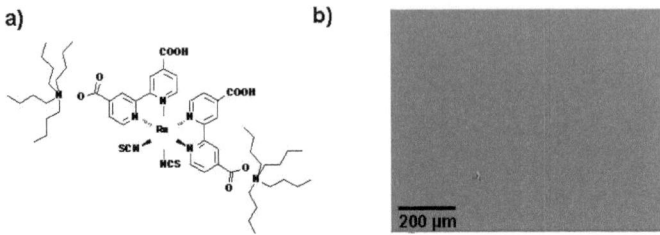

Figure A.1: (a) Ruthenium dye complex deposited on the TiO$_2$ films. (b) SEM picture of the sol-gel TiO$_2$ intralayer surface after calcination at 550 °C.

Solid-state DSSC devices

Compared to the liquid-state solar cells, the solid-state ones are prepared with a different approach based on the presence of a compact TiO$_2$ intralayer between the FTO substrate and the porous TiO$_2$ film. Within this work two different intralayers were compared:

Sputtered TiO$_2$. TiO$_2$ was sputtered on the surface of the FTO glass substrate in order to obtain a compact layer of 15 nm.

Sol-gel TiO$_2$. Compact crystalline non-templated TiO$_2$ intralayer films were prepared following the procedure described in A.1.4. However, in this case the amount of polymer was only 5% wt., i.e. 1.5 mg, in order to facilitate the attachment of the sol on the FTO substrate surface. The samples were dip-coated with a constant withdrawal speed of 0.5 mm/s in order to reach a thickness of ca. 50 nm. The homogeneity and the crack-free surface of the materials can be observed by the SEM picture in Fig. A.1b.

Two different dye sensitized heterojunctions were studied in this work.

Nanoparticulate TiO$_2$-ECN systems. A suspension of ECN-TiO$_2$/terpineol (1:1 wt.) paste was deposited on the substrates coated with sputtered/sol-gel TiO$_2$ and sintered for 30 min at 450 °C. The nanoparticulated TiO$_2$-ECN films were soaked with an ethanolic solution 0.5 mM of the Ru dye for 18 h. The samples were then washed with ethanol and let dry for 2 h at 40 °C. 150 μl of hole transporting material (HTM) solution, developed by BASF AG, was brought on the device by spin-coating for 10 s at 1000 rmp. After washing with toluole, the samples were let dry in oven for 20 h at 40 °C in vacuum. Finally, a mask of gold electrode was deposited on the top layer by chemical vapor deposition (CVD) to create the contacts.

Sol-gel TiO$_2$ systems. The mesoporous TiO$_2$ layer was prepared according to the synthesis described in section A.1.4. The assembling of the device, i.e. incorporation of the Ru dye, HTM solution and deposition of the gold contacts, was achieved following the same procedure as for the Nanoparticulated TiO$_2$-ECN systems.

A.2 Analytical methods

Gas physisorption measurements

The gas physisorption measurements have been performed in an automated gas adsorption station (Autosorb-1-MP, Quantachrome Corporation, Boynton Beach, FL). The device is dedicated to standard characterization measurements of nanostructured matter by nitrogen sorption isotherms at 77 K. The instrument data reduction sofware supports the standard data reduction algorithms like BET, BJH, etc. as well as the newest DFT kernels for typical pore geometries. The system is equipped with a vapor sorption option consisting of a heated manifold and vapor generator, which allows volumetric vapor sorption measurements at temperatures up to 50 °C. For the gas adsorption measurement the samples have been filled in standard glass cuvettes and have been stabilized on the measurement temperature in the liquid nitrogen filled standard dewar at 77 K for nitrogen sorption, and in a water thermostated bath at 290 K and 276 K, for the dibromomethane and perfluoropentane investigations, respectively. The isotherms have been measured up to 0.95 of the equilibrium vapor pressures $p°$ of the different adsorbents (N$_2$: 1013 mbar, CH$_2$Br$_2$: 36 mbar, C$_5$F$_{12}$: 330 mbar). One measurement took about 5 hours for nitrogen and perfluoropentane, and about 15 hours in the case of the dibromomethane. Before every measurement the samples were degased at 120 °C for minimum 6 h.

SAXS measurements

Transmission and low-angle transmission 2D SAXS measurements were performed using a Nonius roatating anode (Cu Kα radiation, $\lambda = 0.154$ nm) and a MAR-CCD area-detector (sample-detector distance of 740 mm). The scattering patterns

A. Appendix

were resolved using the data reduction sotware FIT2D [168]. The mesoporous films were deposited on ultrathin silicon wafers (thickness 40 μm), thus allowing measurements in transmission mode.

WAXS measurements

A D8 Advance machine from Brucker Instruments was used for the WAXS measurements of TiO_2 thin films. The x-ray tube operated at 40 kV and 40 mA, emitted a Cu Kα radiation that was chromatized by a multilayer Göbel mirror. An energy dispersive detector (Sol-X, Brucker) was used to ensure low background noise. The samples were placed in deepened plastic sample holders. The measurements were performed in reflection geometry as θ-2θ scans between 15 ° - 80 °, scanspeed 0.5 sec/step, increment 0.02.

Electron microscopy measurements

High resolution SEM measurements were performed using a Gemini 1550 (120kV, Carl Zeiss, Oberkochen). TEM images were recorded with a Zeiss EM 912 Ω at an acceleration voltage of 120 keV. Regarding the TEM measurements, the samples were dispersed in aceton and a drop of the obtained solution was let dry on a carbon coated copper grid (400 mesh).

AFM measurements

Tapping mode AFM images were recorded with a multimode AFM from Vecco Instruments employing Olympus microcantilevers (resonance frequency, 300 kHz; force constant, 42 N/m).

A.3 In-situ SAXS/SANS-physisorption

A.3.1 In-situ SAXS-physisorption

For the in-situ small-angle x-ray scattering (SAXS) measurements in conjunction with sorption of a fluid in the sample, a special custom-made apparatus was designed. For further details of the setup see reference [153]. The sample powder was carefully pressed to stable pellet of 3 mm diameter and 0.3 mm thickness. Before starting a sorption experiment, the sorption cell was evacuated at a pressure below 10^{-3} mbar at a temperature of 80 °C for one hour, but the sample had been kept under vacuum at 120 °C prior to bringing it into the in-situ cell. For sorption experiments, the cell was thermostated at 17 °C to be the coldest point in the system, while the liquid reservoir stayed at ambient temperature of 25 °C. The liquid adsorptive dibromomethane CH_2Br_2 was used as received. The sorption cell was controlled remotely by a custom-written software program, which allowed continuous adsorption and desorption scans. The SAXS measurements were performed at the beamlines BW4 [169] at the Hamburger Synchrotronstrahlungslabor (HASYLAB)/Deutsches Elektronen-Synchrotron (DESY) in Hamburg, Germany. The synchrotron radiation was monochromatized by a double crystal monochromator to an energy E of 8.27 keV and 8.979 KeV ($\Delta E/E = 10^{-3}$), respectively, focused by a single mirror, and the cross section of the beam was defined by aperture slits to 0.5×0.5 mm^2 at the sample position. For the detection of the scattered photons a charge-coupled device (CCD) x-ray area detector (marCCD 165, marUSA, Evanston, IL, USA) with a resolution of 2048×2048 pixels (pixel size 79.1×79.1 μm^2) was used. The precise sample-to-detector distance was determined by a calibration with a standard sample of silver behenate [170]. A total range of the scattering vector s of 0.03 nm^{-1} < s < 0.38 nm^{-1} was covered. In order to avoid air scattering a vacuum flight tube was inserted between the sample and the detector. The transmission of the investigated sample at each sorption state was determined in-situ by using an ionization chamber before the adsorption cell for monitoring the primary synchrotron x-ray flux and a photodiode mounted in the beamstop for measuring the transmitted photons. An exposure time of 60 s yielded a scattering pattern with excellent counting statistics. The scattering patterns were corrected for background scattering, electronic noise, transmission and polarization by using the data reduction software program FIT2D [168]. All specimens showed isotropic scattering patterns, which were azimuthally averaged for equal radial distances from the central beam.

A.3.2 In-situ SANS-physisorption

The in-situ SANS experiments shown in this work were carried out at two different beam lines, namely the membrane diffractometer V1 and the small-angle neutron scattering instrument V4, of the Helmholtz Zentrum Berlin, Germany,

A. Appendix

which are here described.

Membrane Diffractometer V1. The diffractometer is located at the curved neutron guide NL 1A. The vertically focusing graphite monochromator provides adjustable wavelengths between 0.3 nm and 0.6 nm, making use of the full cross section of the neutron guide (3 cm wide, 12.5 cm high). A movable Be-filter at liquid nitrogen temperature was used to suppress second order wavelengths below 0.39 nm. If less stringent collimation conditions allow, high flux at short monochromator-sample distances (approx. 80 cm) are made possible by a compact construction of shieldings. Sample and detector supports are movable on aircushions. The maximum scattering vector s value in all configurations with Be-Filter is limited to 3.5 nm^{-1}. The in situ SANS measurements in this work were performed in the range 0.00334 nm^{-1} < s < 1.12 nm^{-1}. The ^{3}He-detector provides a sensitive area of 19×19 cm^2 with a spatial resolution of 1.5 x 1.5 mm^2. Instrument control and data acquisition is implemented by CAMAC interfaced to a Compaq Alpha Workstation using the instrument control program CARESS.

Small-Angle Neutron Scattering (SANS) Instrument V4. The instrument is installed at the curved neutron guide NL3a. Incoming neutrons are monochromatized by a mechanical velocity selector with variable wavelength resolution (from 8% to 18%) and collimated on a variable length from 1 to 16 m. The two-dimensional position sensitive detector can be positioned at any distance between 1 m and 16 m from the sample in the horizontal direction. Within this work the measurements were performed in the range 0.002 nm^{-1} < s < 0.5 nm^{-1}, applying sample-detector distances of 1.4 m, 4 m and 16 m. The ^{3}He detector of 64×64 cm^2 is equipped with new read-out electronics for 128×128 cells. A large sample chamber is connected to the vacuum system with the detector and collimator tubes. The instrument is fully controlled via CAMAC by an ALPHA workstation using the instrument control program CARESS. The acquired scattering patterns were corrected by background and noise effects with the help of calibration measurements of water and Cd.

In the in-situ SANS experiments the relative pressure $p/p°$ of the adsorptive could be precisely controlled. Before performing each scattering measurement the sample material was filled with a certain amount $(n/n°)$ of condensate by using an appropriate gas adsorption sample environment (CGA-PT) which allows a direct in-situ measurement of an entire pV isotherm. The porous material was situated in an aluminum cylindrical cell mounted on the cold finger of a closed cycle refrigerator stabilized at the desired temperature.

A.4 Mathematical appendix

A.4.1 Expressions of $F(s)$ and $S(s)$ for hard-spheres system in the PY model

The two contribution of the form factor ($F(s)$) of a sphere used in the Percus-Yevick algorithm are

$$\langle F(s) \rangle = \frac{H(s/2)}{2\pi s^3}[(1 + 4\pi^2 \sigma_R) \sin 2\pi \langle R \rangle s - 2\pi \langle R \rangle s \cos 2\pi \langle R \rangle s] \quad (A.1)$$

$$\langle F(s)^2 \rangle = \frac{1}{8\pi^4 s^6} + \frac{\sigma_R^2 + \langle R^2 \rangle}{2\pi^4 s^6} \left\{ 1 + H(s) \left[\left(1 - \frac{1}{4\pi^2 s^2 (\sigma_R^2 + \langle R^2 \rangle)} \right.\right.\right.$$
$$\left.\left. - \frac{4\sigma_R^2}{\sigma_R^2 + \langle R^2 \rangle}(1 + 4\pi^2 \sigma_R^2 s^2) \right) \cos 4\pi \langle R \rangle s \right.$$
$$\left.\left. - \frac{\langle R \rangle}{\sigma_R^2 + \langle R^2 \rangle} \left(8\pi \sigma_R^2 s + \frac{1}{\pi s}\right) \sin 4\pi \langle R \rangle s \right] \right\} \quad (A.2)$$

where $H(s) = \mathscr{F}(h(r)) = \exp(-4\pi^2 \sigma_R^2 s^2)$.

The correlation function of Orstein, employed for the determination of the structure factor ($S(s)$) is given by

$$G(A) = \alpha \frac{\sin A - A \cos A}{A^2} + \beta \frac{2A \sin A + (2 - A^2) \cos A - 2}{A^3}$$
$$+ \gamma \frac{-A^4 \cos A + 4[(3 - A^2 - 6) \cos A + (A^3 - 6A) \sin A + 6]}{A^5} \quad (A.3)$$

where α, β, γ are functions of η_{PY}

$$\alpha = (1 + 2\eta_{PY})^2/(1 - \eta_{PY})^4$$
$$\beta = -6\eta_{PY}(1 + \eta_{PY}/2)^2/(1 - \eta_{PY})^2 \quad (A.4)$$
$$\gamma = \eta_{PY} \alpha/2$$

A. Appendix

A.4.2 Expressions of $F(s)$ and $S(s)$ for hard-discs system in the PY model

As in the case of spherical systems, the analytical calculation of the scattering intensity assuming a polydispersed hard-discs model in the Percus-Yevick approximation and considering that the position of the cylindrical mesopores is independent from their size, is given by Eqn. 1.24.

The form factor of a cylinder of radius R is given by

$$F(R,s) = \frac{R}{s} J_1(2\pi Rs) \qquad (A.5)$$

where the $J_1(R,s)$ is a Bessel function of the first kind of order 1. The two contributions of the form factor in Eqn. 1.24 can be expressed by

$$\langle F(R,\sigma_R,s)\rangle = \frac{R^2 + \sigma_R^2}{R^2} \exp(-2\pi^2 s^2 \sigma_R^2) F(R,s) \qquad (A.6)$$

$$\langle F(R,\sigma_R,s)^2\rangle = \frac{2}{2\pi^2 s^3}\left[\exp(-8\pi^2 s^2 \sigma_R^2) \cdot \right.$$
$$\left. \left(\frac{2\pi^2 s^3 (R^4 + 6R^2\sigma_R^2 + 3\sigma_R^4)F(R,s)}{R^5} - 1\right) + 1\right] \qquad (A.7)$$

The structure factor $S(s)$ is obtained in the Percus-Yevick approximation through the Rosenfeld approach [167] with the parameter η_{PY} and R_{PY}

$$S(\eta_{PY}, R_{PY}, s) = \left\{4\eta_{PY}\left[A(\eta_{PY})\left(\frac{J_1(R_{PY}s)}{R_{PY}s}\right)^2 + \frac{B(\eta_{PY})J_0(R_{PY}s)J_1(R_{PY}s)}{R_{PY}s}\right.\right.$$
$$\left.\left. + \frac{G(\eta_{PY})J_1(2R_{PY}s)}{R_{PY}s}\right] + 1\right\}^{-1} \qquad (A.8)$$

$$A(\eta_{PY}) = \frac{1 + (2\eta_{PY} - 1)\chi(\eta_{PY}) + 2\eta_{PY}G(\eta_{PY})}{\eta_{PY}}$$

$$B(\eta_{PY}) = \frac{(1 - \eta_{PY})\chi\eta_{PY} - 1 - 3\eta_{PY}G(\eta_{PY})}{\eta_{PY}}$$

$$G(\eta_{PY}) = \frac{1}{(1 - \eta_{PY})^3} \quad ; \quad \chi(\eta_{PY}) = \frac{1 + \eta_{PY}}{(1 + \eta_{PY})^3}$$

where J_0 and J_1 are the Bessel functions having the property that $J_0(x)$ and $2J_1(x)/x$ are both equal to 1 for $x = 0$.

A.4.3 Calculation of the $\breve{\phi}_{micro}$ from the Bragg peak intensity

The Eqn. 3.3 can be obtained as it follows. The scattering length density (SLD) of the amorphous silica is defined as $\rho_{b_{SiO_2}}$, according to an average mass density of 2.2 g/cm^3, and as $\breve{\rho}_{b_{void}}$ the averaged scattering length density of the silica matrix between the block copolymer pores in the evacuated state. Assuming that the SLDs of silica and liquid nitrogen are practically identical, hence it can be defined $\breve{\rho}_{b_{micro}}$ as the SLD of the matrix between the block copolymer pores with the micropores being filled. Accordingly, $\breve{\rho}_{b_{IL}}$ is the scattering length density of the matrix between the block copolymer pores with *also* the small IL mesopores filled. At the void state the relation holds true

$$\breve{\rho}_{b_{void}} = \rho_{b_{SiO_2}}(1 - \breve{\phi}_{IL} - \breve{\phi}_{micro})$$
$$\approx \rho_{b,SiO_2}(1 - \breve{\phi}_{IL})(1 - \breve{\phi}_{micro})$$

The latter approximation is valid because the product $\breve{\phi}_{IL}\breve{\phi}_{micro}$ is rather small. This adjustment allows a further development of a relationship between the intensity of the Bragg peak and the SLDs, as follows

Assuming

$$\breve{\rho}_{b_{micro}} = \breve{\rho}_{b_{SiO_2}}(1 - \breve{\phi}_{IL})$$

then

$$\breve{\rho}_{b_{void}} = \breve{\rho}_{b_{micro}}(1 - \breve{\phi}_{micro})$$

From the increase in intensity of the Bragg peak maximum

$$\frac{I_{void}}{I_{micro}} = \frac{\breve{\rho}_{b_{void}}^2}{\breve{\rho}_{b_{micro}}^2} = k$$

then

$$\frac{\breve{\rho}_{b_{void}}}{\breve{\rho}_{b_{micro}}} = \sqrt{k}$$

Consequently it follows

$$(1 - \breve{\phi}_{micro}) = \sqrt{k} = \sqrt{\frac{I_{void}(s)}{I_{micro}(s)}} \qquad (A.9)$$

A. Appendix

A.5 Discussions on micropore analysis

A.5.1 Monolayer formation of adsorbate in hierarchical SiO$_2$

The following analysis is realized considering the KLE-IL system.

Block copolymer mesopores. The influence of the layer formation within block copolymer mesopores on the SANS at small pressures ($p/p° < 0.01$) is negligible, as demonstrated by a straightforward calculation. Assuming $r = 7$ nm for the block copolymer mesopore radius and assuming that a layer of a thickness of 0.3 nm forms at $p/p° = 0.008$ (which is an exaggeration), the reduced mesopore radius is $r = 6.7$ nm (Fig. A.2). Comparing the form factor of spheres for these two radii and using a polydispersity of 5% for the size distribution (Gauss distribution), differences are seen in the minima. However, the two curves are practically identical at small scattering vectors, i.e. in the region of the Bragg peak. In particular, the absolute scattering intensities are practically identical, so that the analysis of the Bragg reflection is practically not disturbed by layer formation at small pressures of ca. $p/p° < 0.01$.

IL mesopores. Assuming cylindrical IL mesopores of a radius of 1.5 nm and layer thickness of 0.2 nm of nitrogen (which is a rough number) for $p/p° = 0.008$, the volume fraction of the layer with respect to the total volume of an IL mesopore is ca. 0.25. Assuming a layer thickness of 0.3 nm at $p/p° = 0.008$ as upper limit, the volume occupied by the "layer" is about 30% of the volume of an IL mesopore. (Such estimation is not exact, of course, since the IL mesopores are not ideally cylindrical, but gives good approximation). Taking the cumulative IL mesopore volume measured by physisorption (0.18 cm^3/g, Tab. 3.1), thus maximum 25%-30% of this value, which is ca. at maximum 0.04 cm^3/g, correspond reasonably to the difference in the micropore volume determined from NLDFT and the SANS analysis (0.07 cm^3/g). While this calculation is just an approximation, it demonstrates that the layer formation has to be taken into account, and that the micropore content can be determined with reasonable accuracy. However, it cannot explain the increase in intensity in the Bragg peak, thus micropores must be present.

A.5.2. Volume fraction calculation by means of the Bragg peak analysis

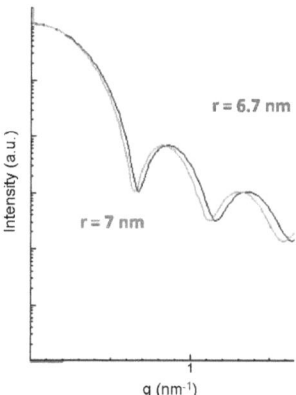

Figure A.2: Simulated scattering patterns of the form factor of a sphere having different radius r.

A.5.2 Volume fraction calculation by means of the Bragg peak analysis

The different volume fractions ϕ were obtained by means of the Eqn. 3.2. In the case of the NLDFT method the single volumina were taken from Tab. 3.1. The application of the Bragg peak analysis considers the determination from the scattering pattern of the "partial" volume fractions $\breve{\phi}$, i.e. the volume fractions of the pores (micropores and small IL mesopores) in the walls separating the block copolymer mesopores. Through the conversion of the Eqn 3.1 and similar, the single volumina values can be obtained as follows:

$$V_{IL+micro} = \frac{\breve{\phi}_{IL+micro} \, V_{SiO_2}}{1 - \breve{\phi}_{IL+micro}} \quad (A.10)$$

$$V_{micro} = (1 - \breve{\phi}_{IL+micro}) \, \breve{\phi}_{micro} \, V_{SiO_2} \quad (A.11)$$

where V_{SiO_2} is obtained by the density of silica ρ_{SiO_2} (2.2 cm^3/g) and the mass of the analysed sample in Tab. 3.1. The value of V_{IL} is obtained by the difference between $V_{IL+micro}$ and V_{micro}. Since V_{BC} pores can not be determined with the Bragg peak analysis, the value obtained with the NLDFT is used also within this approach. The practical example of the ϕ_{micro} calculation for KLE-IL silica is

A. Appendix

shown here below: From Eqn. 3.3

$$1-\breve{\phi}_{IL+micro}=\sqrt{\frac{27}{54}}=0.7$$

$$\breve{\phi}_{IL+micro}=0.3$$

$$1-\breve{\phi}_{micro}=\sqrt{\frac{27}{39}}=0.83$$

$$\breve{\phi}_{micro}=0.17$$

then from Eqn. A.10 and A.11:

$$V_{IL+micro} = \frac{0.3 \times 0.053}{0.7} \; cm^3 = 0.023 \; cm^3$$

$$V_{micro} = 0.7 \times 0.17 \times 0.053 \; cm^3 = 0.0063 \; cm^3$$

$$V_{IL} = V_{IL+micro} - V_{micro}$$

$$= (0.023 - 0.063) \; cm^3 = 0.0167 \; cm^3$$

From Tab. 3.1 the value of $V_{BC} = 0.0151$ cm^3 for KLE-IL is obtained, then with Eqn. 3.2

$$\phi_{micro} = \frac{0.0063 \; cm^3}{(0.0063 + 0.0167 + 0.0151 + 0.053) \; cm^3} = 0.07$$

A.6 Additional physisorption analyses

A.6.1 N$_2$ physisorption on different batch of PIB-IL

The present batch of PIB-IL hierarchical silica was used in the vapor physisorption and in the in-situ SAXS/SANS-vapor physisorption studies.

As one can see in Fig. A.3 the porous and morphological features of the material are in good agreement with those of the batch investigated in chapter 3, cf. Fig. 3.2 and Tab. 3.1, thus proving the good reproducibility of the synthesis procedure. The BET surface area was measured in the range $0.05 \leq p/p^\circ \leq 0.15$ and is found to be of 788 m^2/g.

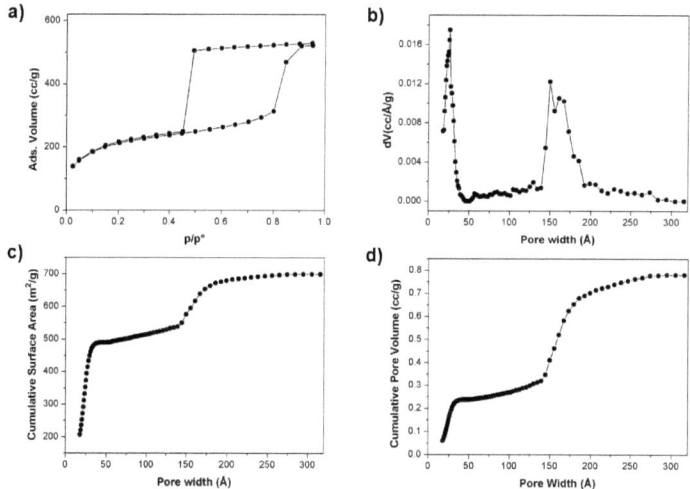

Figure A.3: (a) Nitrogen physisorption isotherm of a second batch of PIB-IL. NLDFT pore size distribution (b), cumulative surface area (c) and cumulative pore volume (d) calculated from the adsorption branch of the physisorption isotherm by applying the kernel of metastable adsorption isotherms based on a spherical/cylindrical pore model for the system nitrogen (77.4 K)/silica.

A.6.2 Langmuir plot of CH_2Br_2 physisorption on PMOs

In Fig. A.4 the Langmuir plots of the CH_2Br_2 isotherms of both PMO materials are depicted. It can clearly be seen, that the lower correlation coeffcient with respect to the BET plot one (cf. Fig. 5.4) excludes Langmuir as correct physisorption behavior.

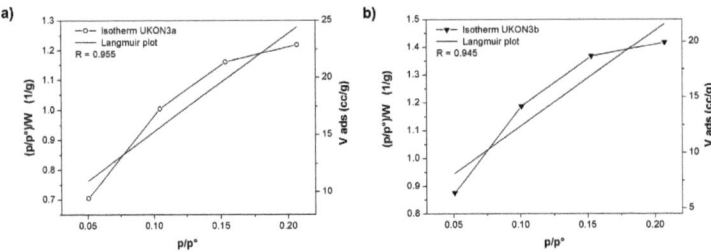

Figure A.4: Langmuir plot of the CH_2Br_2 isotherms of UKON3a (a) and UKON3b (b) in the range $0.05 \leq p/p° \leq 0.2$. The term W in the left y-axis express the mass of adsorbed CH_2Br_2, R represents the correlation coefficient.

A.7 List of Chemicals

C_5F_{12}	Perfluoropentane (97%), Alfa Aesar
CH_2Br_2	Dibromomethane ($\geq$ 98.5%), Fluka
CH_3CH_2OH	Ethanol absolute ($\geq$ 99.5%), Normapur
PIB2300	Granted by BASF AG
PIB6000	Granted by BASF AG
TEOS	Tetraethoxysilane ($\geq$ 99.0%), Sigma Aldrich
THF	Tetrahydrofuran ($\geq$ 99.0%), Sigma Aldrich
$TiCl_4$	Titanium tetrachloride ($\geq$ 99.0%), Fluka
TMOS	Tetramethoxysilane ($\geq$ 99.0%), Sigma Aldrich

Acronyms and Abbreviations

ABC	Amphiphilic Block Copolymer
AFM	Atomic Force Microscopy
BCC	Body Centered Cubic
BET	Brunauer Emmet Teller method
BJH	Barret Joyner Halenda method
CLD	Chord Length Distribution
CMC	Critical Micelle Concentration
DFT	Density Functional Theory
DSSC	Dye Sensitized Solar Cell
EISA	Evaporation-Induced Self-Assembly
FCC	Face Centered Cubic
FF	Fill Factor
FTO	Fluorine Tin Oxide
GISAXS	Grazing Incidence Small-Angle X-ray Scattering
HCP	Hexagonal Close Packed
HTM	Hole-Transport Material
IL	Ionic Liquid
IUPAC	International Union of Pure and Applied Chemistry
KL	Kraton Liquid
KLE	Kraton Liquid poly(Ethylene oxide)
NLDFT	Non-Local Density Functional Theory
PEO	Poly(Ethylene Oxide)
PIB	Polyisobutylene
PMO	Periodic Mesoporous Organosilica
PPO	Poly(Propylene Oxide)
PSD	Pore Size Distribtion
PY	Percus-Yevick

Acronyms and Abbreviations

QE	Quantum Efficiency
SANS	Small-Angle Neutron Scattering
SAXS	Small-Angle X-ray Scattering
SE	Secondary Electrons
SEM	Scanning Electron Microscopy
SLD	Scattering Length Density
TCO	Transparent Conductive Oxide
TEM	Transmission Electron Microscopy
TEOS	Tetraethoxysilane
THF	Tetrahydrofuran
TMOS	Tetramethoxysilane
WAXS	Wide-Angle X-ray Scattering

Symbols

a		lattice parameter
A		scattering amplitude
$\bar{a}$		area per hydrophilic group
a_m		molecular cross sectional area of the adsorbate
β		incidence angle
c		emptying factor
C		adsorbate-adsorbent interaction energy
γ		correlation function
γ		surface tension of the bulk fluid
D_{in}		internal pore diameter
F		form factor
g		chord length distribution function
η		density fluctuation
η_{CE}		energy conversion efficiency
η_{PY}		Percus-Yevick volume fraction
h		gaussian distribution of the polydispersity
hkl		Miller index
θ		scattering angle
I		scattering intensity
J		Bessel function
J_{sc}		short-cut current density
λ		wavelength
l		length of hydrocarbon chain
l_p		Porod length
m		mass
n_a		amount of adsorbate
N_A		Avogadro constant

Symbols

n_m	monolayer capacity
N_V^{ex}	kernel of theoretical isotherms in pores of different size
P	Patterson function
p/p°	relative pressure
p_d	pressure of desorption
p_{cp}	pressure at the closure point of hysteresis
q	scattering vector
Q	Porod invariant
r	vector space
R	pore radius
R_{PY}	Percus-Yevick radius
ρ	mass density
ρ_b	scattering length density
$\breve{\rho}_b$	partial scattering length density
ρ_e	electron density
s	scattering vector $(s = q/2\pi)$
S	lattice factor
S	surface area
S_{BET}	BET surface area
σ_R	variance of the pore radius
T	temperature
t_c	thickness of an adsorbed multilayer film
v	volume occupied by the surfactant
V	volume
V_{cp}	volume at the closure point of hysteresis
V_d	volume of desorption
V_{OC}	voltage at opened circuit
V_p	total pore volume
ϕ	volume fraction
Φ	volume fraction of the block copolymer chains
$\breve{\phi}$	partial volume fraction
φ_V	pore size distribution function
χ	chain length fraction
$\otimes$	pore size

Bibliography

[1] Mann, S. *Biomimetic Materials Chemistry* (S. Mann, ed.) **1997**, Wiley-VCH, Weinheim, DE, 1 40.

[2] Calvert, P. *Biomimetic Materials Chemistry* (S. Mann, ed.) **1997**, Wiley-VCH, Weinheim, DE, 315 - 336.

[3] Sarikaya, M.; Aksay, I. *Biomimetics: Design and Processing of Materials* (M. Sarikaya, I. Aksay, eds.) **1995**, AIP, Woodbury, Connecticut, U.S.

[4] Bäuerlein, E. *Angew. Chem. Int. Ed.* **2003**, *42*, 614.

[5] Mann, S.; Burkett, S.L.; Davis, S.A. *Chem. Mater.* **1997**, *9*, 2300.

[6] Ozin, G.A. *Chem. Commun.* **2000**, 419.

[7] Van Bommel, K.J.C.; Friggeri, A.; Shinkai, S. *Angew. Chem. Int. Ed.* **2003**, *42*, 980.

[8] Schüth, F. *Chem. Mater.* **2001**, *13*, 3184.

[9] Tolbert, S.H.; Firouzi, A.; Stucky, G.D.; Chmelka, B.F. *Science* **1997**, *278*, 264.

[10] Sanchez, C.; Arribart, H.; Giraud-Guille, M.M. *Nat. Mater.* **2005**, *4*, 277.

[11] Giraud-Guille, M.M. *Int. Rev. Cytol.* **1996**, *166*, 59.

[12] Peytcheva, A.; Antonietti, M. *Angew. Chem. Int. Ed.* **2001**, *17*, 3380.

[13] Sleytr, U. B.; Schuster, B.; Pum, D. *Eng. Med. Biol. Mag.* **2003**, *22*, 14.

[14] Kenichi, K.; Kazushi, F.; Junzo, S.; Kazunari, A. *Langmuir* **2002**, *18*, 3780.

[15] Mann, S.; Ozin, G.A. *Nature* **1996**, *382*, 313.

[16] Yang, H.; Ozin, G.A.; Kresge, C.T. *Adv. Mater.* **1998**, *10*, 883.

[17] Yang, S.M.; Sokolov, I.; Coombs, N.; Kresge, C.T.; Ozin, G.A. *Adv. Mater.* **1999**, *11*, 1427.

[18] Schüth, F.; Schmidt, W. *Adv. Mater.* **2002**, *14*, 629.

[19] Sing, K.S.W.; Everett, D.H.; W. Haul, R.A.; Moscou, L.; Pierotti, J.; Rouquerol, J.; Siemieniewska, T. *Pure Appl. Chem.* **1985**, *57*, 603.

[20] Sing, K.S.W. *Pure Appl. Chem.* **1982**, *54*, 2201.

[21] Thommes, M. *Nanoporous Materials: Science and Engineering* (G.Q. Lu, X S. Zhao, eds.) **2004**, Imperial College Press London, U.K., 317.

[22] Brunauer, S.; Emmet, P.H.; Teller, E. *J. Am. Chem. Soc.* **1938**, *19*, 309.

[23] Gregg, S.J.; Sing, K.S.W. *Adsorption, Surface Area and Porosity* **1982**, Academic Press, London, U.K.

[24] Rouquerol, F.; Rouquerol, J.; Sing K. S.W. *Adsorption by Powders & Porous Solids* **1999**, Academic Press, London, U.K.

[25] Lowell, S.; Shields, J.E. *Powder Surface Area and Porosity* **1991** Chapman & Hall, London, U.K.

[26] Barrett, E.P.; Joyner, L.G.; Halenda, P.P. *J. Am. Chem. Soc.* **1951**, *73*, 373.

[27] Ravikovitch, P.I.; O'Domhnaill, S.C.; Neimark, A.V.; Schüth, F.; Unger, K.K. *Langmuir* **1995**, *11*, 4765.

[28] Ravikovitch, P.I.; Wei, D.; Chueh, W.T.; Haller, G.L.; Neimark, A.V. *J. Phys. Chem. B* **1997**, *101*, 3671.

[29] Neimark, A.V.; Ravikovitch, P.I. *Micro. Meso. Mater.* **2001**, *44-45*, 697.

[30] Neimark, A.V.; Ravikovitch, P.I.; Vishnyakov, A. *Phys. Rev. E* **2000**, *62*, R1493.

[31] Ravikovitch, P.I.; Neimark, A.V. *Langmuir* **2002**, *18*, 1550.

[32] Ravikovitch, P.I.; Neimark, A.V. *Colloids Surf., A* **2001**, *11*, 187.

[33] Ravikovitch, P.I.; Neimark, A.V. *Langmuir* **2002**, *18*, 9830.

[34] Thommes, M.; Smarsly, B.; Groenewolt, M.; Ravikovitch, P.I.; Neimark, A.V. *Langmuir* **2006**, *22*, 756.

[35] Sarkisov, L; Monson, P.A. *Langmuir* **2001**, *17*, 7600.

[36] Schofield, R. K. *Discuss. Faraday Soc.* **1948**, *3*, 105.

[37] Kadlec, O.; Dubinin, M.M. *J. Coll. Interface Sci.* **1969**, *31*, 479.

[38] Burgess, C.G.V.; Everett, D.H. *J. Colloid Interface Sci.* **1970**, *33*, 611.

[39] Glatter, O.; Kratky, O. *Small Angle X-ray Scattering* **1982**, Academic Press, London, U.K.

[40] Guinier, A.; Fournet, G. *Small-angle Scattering of X-rays* **1955**, Wiley, New York, U.S.

[41] Porod, G. *Small-angle X-ray Scattering* **1965**, H. Brumberger, ed., 1.

[42] Porod, G. *Kolloid-Z.* **1951**, *124*, 83.

[43] Pedersen, S.J. *Phys. Rev. B* **1993**, *47*, 657.

[44] Vrij, A. *J. Chem. Phys.* **1979**, *71*, 3267.

[45] a) Percus, J.K.; Yevick, G.J. *Phys. Rev.* **1958**, *1*, 110. b) Percus, J.K. *Phys. Rev.* **1962**, *8*, 462.

[46] Ashcroft, N.W.; Lekner, J. *Phys. Rev.* **1966**, *145*, 83.

[47] Siemann, U.; Ruland, W. *Colloid Polym. Sci.* **1982**, *260*, 999.

[48] Kotlarchyk, M.; Chen, S.H. *J. Chem. Phys.* **1983**, *79*, 2461.

[49] Pedersen, J.S. *J. Appl. Cryst.* **1994**, *27*, 595.

[50] Kinning, D.J.; Thomas, E.L. *Macromolecules* **1984**, *17*, 1712.

[51] Wertheim, M.S. *Phys. Rev. Lett.* **1963**, *10*, 321.

[52] Méring, J.; Tchoubar, C. *J. Appl. Crystallogr.* **1968**, *1*, 153.

[53] Tchoubar-Vallat, D.; Méring, J. *C. R. Hebd. Seances Acad. Sci.* **1965**, *261*, 3096.

[54] Méring, J.; Tchoubar-Vallat, D. *C. R. Acad. Sci. Paris* **1966**, *262*, 1703.

[55] Ciccariello, S.; Cocco, G.; Benedetti, A.; Enzo, S. *Phys. Rev. B* **1981**, *23*, 6474.

[56] Ciccariello, S.; Benedetti, A. *Phys. Rev. B* **1982**, *26*, 6384.

[57] Ciccariello, S. *Acta Crystallogr., Sect. A: Found. Crystallogr.* **1993**, *49*, 398.

[58] Ciccariello, S.; Sobry, R. *Acta Crystallogr., Sect. A: Found. Crystallogr.* **1995**, *51*, 60.

[59] Smarsly, B.; Antonietti, M.; Wolff, T. *J. Chem. Phys.* **2002**, *116*, 2618.

[60] Burger, C.; Ruland, W. *Acta Crystallogr., Sect. A: Found. Crystallogr.* **2001**, *49*, 6381.

[61] Engl, H.W.; Hanke, M.; Neubauer, A. *Regularization of Inverse Problems* **1996**, Kluwer, Dordrecht, NED

[62] Hanse, P.C. *Rank-Deficient and Disrecete III-Posed Problems* **1998**, SIAM Monographs, Philadelphia, U.S.

[63] Morozov, V.A. *Methods for Solving Incorrectly Posed Problems* **1984**, Springer-Verlag, New York, U.S.

[64] Ruland, W.; Smarsly, B. *J. Appl. Crystallogr.* **2007**, *40*, 409.

[65] Grosso, D.; Soler-Illia, G.J.A.A.; Crepaldi, E.L.; Cagnol, F.; Sinturel, C.; Bourgeois, A.; Brunet-Bruneau, A.; Amentisch, H.; Albouy, P.A.; Sanchez, C. *Chem. Mater.* **2003**, *15*, 4562.

[66] Ramsay, J.D.F. *Adv. Colloid Interface Sci.* **1998**, *76-77*, 13.

[67] Lilov, S.K. *Cryst. Res. Technol.* **1986**, *21*, 1299.

[68] Steriotis, Th.; Mitropoulos, A.; Kanellopoulos, N.; Keiderling, U.; Wiedenmann, A. *Physica B* **1997**, *234*, 1016.

[69] Mitropoulos, A.Ch.; Haynes, J.M.; Richardson, R.M.; Kanellopoulos, N.K. *Phys. Rev. B* **1995**, *52*, 10 035.

[70] Katsaros, F.; Makri, P.; Mitropoulos, A.; Kanellopoulos, N.; Keiderling, U.; Wiedenmann, A. *Physica B* **1997**, *234*, 402.

[71] Ramsay, J.D.F.; Hoinkis, E. *Physica B* **1998**, *248*, 322.

[72] Hoinkis, E. *Langmuir* **1996**, *12*, 4299.

[73] Hoinkis, E. *Adv. Colloid Interface Sci.* **1998**, *77*, 39.

[74] Hoinkis, E.; Allen, A.J. *J. Colloid Interface Sci.* **1991**, *145*, 540.

[75] Thieme Chemistry, *Roempp Chemie Lexikon* **2006**, Verlag, KG, DE.

[76] Glauter, A.M. *Electron Diffraction: An Introduction for Biologists* **1987**, *12*, Elsevier.

[77] Blanchard, C.R. *The Chemical Educator* **1996**, *1*, Springer-Verlag, New York, U.S.

[78] Binnig, G.; Quate, C.F.; Gerber C. *Phys. Rev. Lett.* **1986**, *56*, 930.

[79] Soler-Illia, G.J.A.A.; Sanchez, C.; Lebeau, B.; Patarin, J. *Chem. Rev.* **2002**, *102*, 4093.

[80] Kresge, C.T.; Leonowicz, M.E.; Roth, W.J.; Vartuli, J.C.; Beck, J.S. *Nature* **1992**, *359*, 710.

[81] Beck, J.S.; Vartuli, J.C.; Roth, W.J.; Leonowicz, M.E.; Kresge, C.T.; Schmitt, K.D.; Chu, C.T.W.; Olson, D.H.; Sheppard, E.W.; McCullen, S.B.; Higgins, J.B.; Schlenker, J.L. *J. Am. Chem. Soc.* **1992**, *114*, 10834.

[82] Nagarajan, R.; Ganesh, K. *J. Chem. Phys.* **1989**, *90*, 5843.

[83] Israelachvili, J.N.; Mitchell, J.D.; Ninham, B.W. *J. Chem. Soc. Faraday Trans. 2* **1976**, *72*, 1525.

[84] Brinker, C.J.; Lu, Y.; Sellinger, A.; Fan, H. *Adv. Mater.* **1999**, *11*, 579.

[85] Bosc, F.; Ayral, A.; Albouy, P.A.; Guizard, C. *Chem. Mater.* **2003**, *15*, 2463.

[86] Desphande, A.S.; Pinna, N.; Smarsly, B.; Antonietti, M.; Niederberger, M. *Small* **2005**, *1*, 313.

[87] Kümmel, M.; Grosso, D.; Boissière, C.; Smarsly, B.; Brezesinski, T.; Albuoy, P.A.; Amentisch, H.; Sanchez, C. *Angew. Chem. Int. Ed.* **2005**, *441*, 4589.

[88] Brezesinski, T.; Smarsly, B.; Iimura, K.I.; Grosso, D.; Boissière, C.; Amentisch, H.; Antonietti, M.; Sanchez, C. *Small* **2005**, *1*, 889.

[89] Hagfeld, A.; Grätzel, M. *Chem. Rev.* **1995**, *95*, 49.

[90] Mora-Seró, I. *J. Phys. Chem. B* **2005**, *109*, 3371.

[91] Brezesinski, T.; Fischer, A.; Iimura, K.I.; Sanchez, C.; Grosso, D.; Antonietti, M.; Smarsly, B. *Adv. Funct. Mater.* **2006**, *16*, 1433.

[92] Zhou, Y.; Antonietti, M. *Chem. Commun.* **2003**, 2564.

[93] Newalkar, B.L.; Katsuki, H.; Komarneni, S. *Micro. Meso. Mater.* **2004**, *73*, 161.

[94] Antonietti, M.; Berton, B.; Göltner, C.; Hentze, H.P. *Adv. Mater.* **1998**, *10*, 154.

[95] Rolison, D.R. *Science* **2003**, *299*, 1698.

[96] Tanev, P.T.; Chibwe, M.; Pinnavaia, T.J. *Nature* **1994**, *368*, 321.

[97] Tanev, P.T.; Pinnavaia, T.J. *Science* **1995**, *267*, 865.

[98] Brinker, C.J.; Scherer, G.W. *Sol-Gel Science: The Physics and Chemistry of Sol-Gel Processing* **1990**, 1st ed.; Academic Press Inc.: New York, U.S.

[99] a) Göltner, C.G.; Henke, S.; Weissenberger, M.C.; Antonietti, M. *Angew. Chem. Int. Ed.* **1998**, *37*, 613. b) Göltner, C.G.; Antonietti, M. *Adv. Mater.* **1997**, *9*, 431. c) Weissenberger, M.C.; Göltner, C.G.; Antonietti, M. *Ber. Bunsen-Ges. Phys. Chem.* **1997**, *101*, 1679.

[100] Kramer, E.; Forster, S.; Göltner C.G.; Antonietti, M. *Langmuir* **1998**, *14*, 2027.

[101] a) Zhao, D.; Hou, Q.; Feng, J.; Chmelka, B.F.; Stucky, G.D. *J. Am. Chem. Soc.* **1998**, *120*, 6024. b) Zhao, D.; Feng, J.; Hou, Q.; Melosh, N.; Fredrickson, G.H.; Chmelka, B.F.; Stucky, G.D. *Science* **1998**, *279*, 548.

[102] Thomas, A.; Schlaad, H.; Smarsly B.; Antonietti, M. *Langmuir* **2003**, *19*, 4455.

[103] Smarsly, B.; Grosso, D.; Brezesinski T.; Pinna, N.; Boissière, C.; Antonietti, M.; Sanchez, C. *Chem. Mater.* **2004**, *16*, 2948.

[104] Grosso, D.; Boissière, C.; Smarsly, B.; Brezesinski T.; Pinna, N.; Albouy, P.A.; Amenitsch, H.; Antonietti, M.; Sanchez, C. *Nat. Mater.* **2004**, *3*, 787.

[105] Crepaldi, E.L.; Soler-Illia, G.J.A.A.; Grosso, D.; Cagnol, F.; Ribot, F.; Sanchez, C. *J. Am. Chem. Soc.* **2003**, *125*, 9770.

[106] Smarsly, B.; Polarz, S.; Antonietti, M. *J. Phys. Chem. B* **2001**, *105*, 10473.

[107] Ivonava, R.; Alexandridis, P.; Lindman, B. *Colloids Surf., A* **2001**, *183-185*, 41.

[108] a) Li, Y.; Xu, R.; Bloor, D.M.; Holzwarth, J.F.; Wyn-Jones, E. *Langmuir* **2000**, *16*, 10515. b) Li, Y.; Xu, R.; Couderc, S.; Bloor, D.M.; Wyn-Jones, E.; Holzwarth, J.F. *Langmuir* **2001**, *17*, 183. c) Li, Y.; Xu, R.; Couderc, S.; Bloor, D.M.; Holzwarth, J.F.; Wyn-Jones, E. *Langmuir* **2001**, *17*, 5742.

[109] Sel, O.; Kuang, D.B.; Thommes, M.; Smarsly, B. *Langmuir* **2006**, *22*, 2311.

[110] Groenewolt, M.; Antonietti, M.; Polarz, S. *Langmuir* **2004**, *20*, 7811.

[111] Kuang, D.B.; Brezesinski, T.; Smarsly, B. *J. Am. Chem. Soc.* **2004**, *126*, 10534.

[112] Groenewolt, M.; Brezesinski, T.; Schlaad, H.; Antonietti, M.; Groh, P.W.; Iván, B. *Adv. Mater.* **2005**, *17*, 1158.

[113] Pusey, P.N.; Vanmegen, W. *Nature* **1986**, *320*, 340.

[114] Göltner, C.G.; Smarsly, B.; Berton, B.; Antonietti, M. *Chem. Mater.* **2001**, *13*, 1617.

[115] Erdodi, G.; Iván, B. *Chem. Mater.* **2004**, *16*, 959.

[116] Chiu, J.J.; Pine, D.J.; Bishop, S.T.; Chmelka, B.F. *J. Catal.* **2004**, *221*, 400.

[117] Pauly, T.R.; Liu, Y.; Pinnavaia, T.J.; Billinge, S.J.L.; Rieker, T.P. *J. Am. Chem. Soc.* **1999**, *121*, 8835.

[118] Coppens, M.O.; Froment, G.F. *Fractals* **1997**, *5*, 493.

[119] Sel, O.; Sallard, S.; Brezesinski, T.; Rathouský, J.; Dunphy, D.R.; Collord, A.; Smarsly, B. *Adv. Funct. Mater.* **2007**, *17*, 3241.

[120] Förster, S.; Zisenis, M.; Wenz, E.; Antonietti, M. *J. Chem. Phys.* **1996**, *104*, 9956.

[121] Tate, M.P.; Urade, V.N.; Kowalski, J.D.; Wei, T.C.; Hamilton, B.D.; Eggiman, B.W.; Hillhouse, H.W. *J. Phys. Chem. B* **2006**, *110*, 9882.

[122] Eggiman, B.W.; Tate, M.P.; Hillhouse, H.W. *Chem. Mater.* **2006**, *18*, 723.

[123] Bates, F.S.; Schulz, M.F.; Khandpur, A.K.; Förster, S.; Rosedale, J.H.; Almdal, K.; Mortensen, K. *Faraday Discuss.* **1994**, *98*, 7.

[124] Bates, F.S.; Fredrickson, G.H. *Annu. Rev. Phys. Chem.* **1990**, *41*, 525.

[125] Förster, S.; Berton, B.; Hentze, H.P.; Kämer, E.; Antonietti, M.; Linder, P. *Macromolecules* **2001**, *34*, 4610.

[126] Grosso, D.; Balkenende, A.R.; Albouy, P.A.; Lavergne, M.; Mazzerolles, L.; Babonneau, F. *J. Mater. Chem.* **2000**, *10*, 2085.

[127] Sakya, P.; Seddom, J.M.M.; Templer, R.H.; Mirkin, R.J.; Tiddy, G.J.T. *Langmuir* **1997**, *13*, 3706.

[128] Smarsly, B.; Antonietti, M. *Eur. J. Inorg. Chem.* **2006**, 1111.

[129] Bisquert, J.; Cahen, D.; Hodes, G.; Rühle, S.; Zaban, A. *J. Phys. Chem. B* **2004**, *108*, 8106.

[130] O'Regan, B.; Grätzel, M. *Nature* **1991**, *353*, 737.

[131] Hagen, J.; Schaffrath, W.; Otschik, P.; Fink, R.; Bacher, A.; Schmidt, H.W.; Haarer, D. *Synth. Met.* **1997**, *89*, 215.

[132] Bach, U.; Lupo, D.; Comte, P.; Moser, J.E.; Weissörtel, F.; Salbek, J.; Spreitzer, H.; Grätzel, M. *Nature* **1998**, *395*, 583.

[133] Wang, P.; M. Zakeeruddin, S.; Comte, P.; Exnar, I.; Grätzel, M. *J. Am. Chem. Soc.* **2003**, *125*, 1166.

[134] O'Regan, B.; Lenzmann, F.; Muis, R.; Wienke, J. *Chem. Mater.* **2003**, *14*, 5023.

[135] Nogueira, A.F.; Longo, C.; De Paoli, M.A. *Coord. Chem. Rev.* **2004**, *248*, 1455.

[136] Nelson, J. *The Physics of Solar Cells* **2003**, Imperial College Press, London, U.K.

[137] Lancelle-Beltrame, E.; Prené, P.; Boscher, C.; Belleville, P.; Buvat, P.; Lambert, S.; Guillet, F.; Boissièrre C.; Grosso, D.; Sanchez, C. *Chem. Mater.* **2006**, *18*, 6152.

[138] Wang, H.; Oey, C.C.; Djurisic, A.B.; Xie, M.H.; Leung, Y.H.; Man, K.K.Y.C.; Chan, W.K.; Pandey, A.; Nunzi, J.M.; Chui, P.C. *Appl. Phys. Lett.* **2005**, *87*, 023507.

[139] Jaroniec, M.; Krug, M.; Olivier, J.P. *Langmuir* **1999**, *15*, 5410.

[140] Gelb, L.D.; Gubbins, K.E. *Langmuir* **1999**, *154*, 2097.

[141] Melosh, N.A.; Lipic, P.; Bates, F.S.; Wudl, F.; Stucky, G.D.; Fredrickson, G.H.; Chmelka, B.F. *Macromolecules* **1999**, *32*, 4332.

[142] Myazawa, K.; Inagaki, S. *Chem. Commun.* **2000**, 2121.

[143] Jaroniec, M.; Kruk, M.; Ko, C.H.; Ryoo, R. *Chem. Mater.* **2000**, *12*, 1961.

[144] Ryoo, R.; Ko, C.H.; Kruk, M.; Antochshuk, V.; Jaroniec, M. *J. Phys. Chem. B* **2000**, *104*, 11465.

[145] Jun, S.; Joo, S.H.; Ryoo, R.; Kruk, M.; Jaroniec, M.; Liu, Z.; Ohsuna, T.; Terasaki, O. *J. Am. Chem. Soc.* **2000**, *122*, 10712.

[146] De Paul, S.M.; Zwanziger J.W.; Ulrich, R.; Wiesner, U.; Spiess, H.W. *J. Am. Chem. Soc.* **1999**, *121*, 5727.

[147] Lukens, W.W.; Schmidt-Winkel, P.; Zhao, D.Y.; Feng, J.L.; Stucky, G.D. *Langmuir* **1999**, *15*, 5403.

[148] Ravikovitch, P.I.; Neimark, A.V. *J. Phys. Chem. B* **2001**, *105*, 6817.

[149] Ravikovitch, P.I.; Haller, G.L.; Neimark, A.V. *Adv. Colloid Interface Sci.* **1998**, *77*, 203.

[150] Neimark, A.V.; Ravikovitch, P.I.; Grun, M.; Schüth, F.; Unger, K.K. *J. Colloid Interface Sci.* **1998**, *207*, 159.

[151] Smarsly, B.; Göltner, C.; Antonietti, M.; Ruland, W.; Hoinkis, E. *J. Phys. Chem. B* **2001**, *105*, 831.

[152] Sel, O.; Brandt, A.; Wallacher, D.; Thommes, M.; Smarsly, B. *Langmuir* **2007**, *23*, 4724.

[153] Zickler, G.A.; Jähnert, S.; Wagermaier, S.; Funari, S.S.; Findenegg, G.H.; Paris, O. *Phys. Rev. B* **2006**, *73*, 184109.

[154] Hofmann, T.; Wallacher, D.; Huber, P.; Birringer, R.; Knorr, K.; Schreiber, A.; Findenegg, G.H. *Phys. Rev. B* **2005**, *72*, 064122.

[155] Ramsay, J.D.F.; Kallus, S.; Hoinkis, E. *Stud. Surf. Sci. Catal.* **2000**, *128*, 439.

[156] Chujo, Y. *Curr. Opin. Solid State Mater. Sci.* **1996**, *1*, 806.

[157] Asefa, T.; MacLachan, M.J.; Coombs, N.; Ozin, G.A. *Nature* **1999**, *402*, 867.

[158] Inagaki, S.; Guan, S.; Fukushima, Y.; Oshuna, T.; Terasaki, O. *J. Am. Chem. Soc.* **1999**, *121*, 9611.

[159] Inagaki, S.; Guan, S.; Oshuna, T.; Terasaki, O. *Nature* **2002**, *416*, 304.

[160] Hoffman, F.; Corneluis, M.; Morell, J.; Fröba, M. *Angew. Chem. Int. Ed.* **2006**, *45*, 3216.

[161] Fukuoka, A.; Sakamoto, Y.; Guan, S.; Inagaki, S.; Sugimoto, N.; Fukushima, Y.; Hirahara, K.; Iijima, S.; Ichikawa, M. *J. Am. Chem. Soc.* **2001**, *123*, 3373.

[162] Zhang, L.; Zhang, W.; Shi, J.; Hua, Z.; Li, Y.; Yan, J. *Chem. Commun.* **2003**, 210.

[163] Kuschel, A.; Polarz, S. *Adv. Funct. Mater.* **2008**, *18*, 1272.

[164] Kuschel, A.; Polarz, S. *Angew. Chem. Int. Ed.* **2008**, *49*, 9513.

[165] Polarz, S.; Kuschel, A. *Chem. Eur. J.* **2008**, *14*, 9816.

[166] Polarz, S.; Kuschel, A. *Adv. Mater.* **2006**, *18*, 1206.

[167] Rosenfeld, Y. *Phys. Rev. A* **1990**, *42*, 5978.

[168] Hammersley, A.P.; Svensson, S.O.; Hanfland, M.; Fitch, A.;N. Häusermann, D. *High Press. Res.* **1996**, *14*, 235.

[169] Roth, S.V.; Döhrmann, R.; Dommach, M.; Kuhlmann, M.; Kröger, I.; Gehrke, R.; Walter, H.; Schroer, C.; Lengeler, B.; Müller-Buschbaum, P. *Rev. Sci. Instr.* **2006**, *77*, 085106.

[170] Huang, T.C.; Toraya, H.; Blanton, T.N.; Wu, Y. *J. Appl. Cryst.* **1993**, *26*, 180.

List of Figures

1.1	Classification of physisorption isotherms	6
1.2	Desorption mechanisms in mesoporous materials	10
1.3	Scattering representation by two point centers	12
1.4	Chord length for a single particle and two-phase system	14
1.5	Porod plot for the calculation of the CLD	19
1.6	Representation of $g(r)$ for a bimodal porous system	20
1.7	2D SAXS patterns of mesoporous thin films	22
2.1	Synthetic approaches for mesostructured materials	30
2.2	Molecular structures of the PIB-PEO block copolymers	34
2.3	TEM of PIB6000 templated SiO_2 powders	35
2.4	SAXS of PIB6000 templated SiO_2 powders	36
2.5	TEM of PIB2300 templated SiO_2 powders	38
2.6	SAXS of PIB2300 templated SiO_2 powders	39
2.7	N_2 physisorption on PIB-PEO templated SiO_2 powders	40
2.8	TEM of hierarchical mesoporous SiO_2 powders	42
2.9	SAXS of hierarchical mesoporous SiO_2 powders	43
2.10	N_2 physisorption on hierarchical mesoporous SiO_2 powders	44
2.11	Morphological analyses of the surface of PIB-PEO templated SiO_2 films	46
2.12	TEM of PIB-PEO templated SiO_2 films	47
2.13	SAXS of PIB-PEO templated SiO_2 films	48

LIST OF FIGURES

2.14 SAXS structural analyses of PIB-PEO templated SiO_2 films 49

2.15 Low-angle 2D SAXS of PIB6000 templated SiO_2 films 50

2.16 Shrinkage effect in PIB-PEO templated SiO_2 films 52

2.17 Structural analyses of hierarchical mesoporous SiO_2 film . . 53

2.18 Morphological analyses of the surface of PIB-PEO templated TiO_2 films 55

2.19 TEM of PIB6000 templated TiO_2 films 56

2.20 WAXS and SAXS of PIB-PEO templated TiO_2 films 57

2.21 Crystallites size variation in PIB-PEO templated TiO_2 films 57

2.22 SAXS structural analysis in PIB-PEO templated TiO_2 films as function of block copolymer amount 58

2.23 SAXS structural analysis in PIB-PEO templated TiO_2 films as function of annealing temperature 59

2.24 Low-angle 2D SAXS of PIB-PEO templated TiO_2 films . . . 61

2.25 Schematic representation of dye sensitized solar cells 63

2.26 Quantum efficiency and JV curve for an ideal solar cell system 64

2.27 JV curve for a nanoparticulate and sol-gel dye sensitized solar cell 67

3.1 Templating behaviour of the PEO chains of block copolymers during nanocasting 70

3.2 N_2 physisorption analyses on KLE-IL and PIB-IL 73

3.3 SANS curves on the adsorption branch of KLE-IL and PIB-IL 75

3.4 Schematic representation of the gas adsorption mechanism in hierarchical SiO_2 75

3.5 CLD in the micropore filling region of KLE-IL and PIB-IL during N_2 physisorption 77

3.6 Representation of $l_{p,micro}$ and the relative interface area for KLE-IL and PIB-IL 80

LIST OF FIGURES

3.7 CLD of KLE-IL and PIB-IL relative to the block copolymer pores during N_2 physisorption 85

4.1 CH_2Br_2 and C_5F_{12} physisorption isotherms of KLE-IL and PIB-IL 91

4.2 In-situ SAXS/SANS patterns of KLE-IL and PIB-IL during CH_2Br_2 and C_5F_{12} adsorption 92

4.3 CLD of KLE-IL during C_5F_{12} adsorption 94

4.4 Schematic representation of the porous textures of KLE-IL and PIB-IL 95

4.5 In-situ scattering patterns of the desorption process in PIB-IL 97

4.6 Schematic illustration of the desorption mechanism in the cage-like pores of PIB-IL 98

4.7 Plot of $\langle R \rangle$ and η_{PY} as function of the emptying factor in PIB-IL 99

5.1 Potential distribution of the organic moieties in PMOs ... 104

5.2 PMO materials studied within this work 105

5.3 Structural characterization of UKON3a and UKON3b 108

5.4 CH_2Br_2 and C_5F_{12} physisorption isotherms of UKON3a and UKON3b 109

5.5 In-situ SAXS adsorption/desorption curves of CH_2Br_2 in UKON3a and UKON3b 111

5.6 In-situ SANS adsorption/desorption curves of C_5F_{12} on UKON3a 112

5.7 CLD of CH_2Br_2 adsorption on UKON3a and UKON3b ... 113

5.8 CLD of CH_2Br_2 desorption on UKON3a and UKON3b ... 114

5.9 Relative interface area during CH_2Br_2 adsorption and PY fitting of UKON3a and UKON3b 115

A.1 Ru dye and SEM picture of the sol-gel TiO_2 intralayer used in dye sensitized solar cells 126

LIST OF FIGURES

A.2 Simulated scattering patterns of the form factor of a sphere having different radius r 135

A.3 N_2 physisorption analysis of a second batch of PIB-IL 137

A.4 Langmuir plot of the CH_2Br_2 isotherms on UKON3a and UKON3b. 138

List of Tables

1.1	Features of fluids realizing contrast matching with silica	24
2.1	Pore size values of PIB-PEO templated SiO_2 powders	37
2.2	Pore volumes of PIB-PEO templated SiO_2 powders	41
2.3	TEM pore size values of PIB-PEO templated SiO_2 films	47
2.4	Properties of the liquid- and solid-state solar cells studied	66
3.1	Porosity values of KLE-IL and PIB-IL	74
3.2	Porous fractions of KLE-IL and PIB-IL	79
3.3	Theoretical and experimental volume fractions of the PEO blocks	86
4.1	Pore volumes of KLE-IL and PIB-IL from the C_5F_{12}, CH_2Br_2 and N_2 isotherms	91
5.1	Porosity values of UKON3a and UKON3b	107
5.2	Percus-Yevick analysis during CH_2Br_2 adsorption on UKON3a and UKON3b	116
A.1	Experimental composition of the PIB-PEO templated SiO_2 powders	124
A.2	Template amounts used for the synthesis of SiO_2 and TiO_2 films	125

I want morebooks!

Buy your books fast and straightforward online - at one of world's fastest growing online book stores! Environmentally sound due to Print-on-Demand technologies.

Buy your books online at
www.morebooks.shop

Kaufen Sie Ihre Bücher schnell und unkompliziert online – auf einer der am schnellsten wachsenden Buchhandelsplattformen weltweit! Dank Print-On-Demand umwelt- und ressourcenschonend produziert.

Bücher schneller online kaufen
www.morebooks.shop

KS OmniScriptum Publishing
Brivibas gatve 197
LV-1039 Riga, Latvia
Telefax: +371 686 204 55

info@omniscriptum.com
www.omniscriptum.com

Printed by Books on Demand GmbH, Norderstedt / Germany